B.N. Mandal

Indian Statistical Institute, Calcutta

and

Nanigopal Mandal

R K Mission Residential College, West Bengal

Advances in dual integral equations

CHAPMAN & HALL/CRC

Boca Raton   London   New York   Washington, D.C.

Library of Congress Cataloging-in-Publication Data

Catalog record is available from the Library of Congress.

Contents

Preface

This book addresses state-of-the-art dual integral equations and their many applications to physical problems. The technique of dual integral equations is mainly utilized for successful handling of mixed boundary value problems arising in mathematical physics. A typical boundary value problem of mixed type would be one in which the boundary condition on a part of the boundary surface is given in one form while on the remaining part of this surface it is given in some other form. Dual integral equations arise in a natural way in the course of solving a mixed boundary value problem. A fairly extensive account of dual integral equations with kernels as Bessel functions, trigonometric functions and Legendre functions are available in some chapters in the books by Sneddon (1966, 1972) and Sneddon and Lowengrub (1969). A large number of researchers contributed significantly to the methodology of solution as well as applications of dual integral equations with other special functions as kernel, which are mostly available in the form of research papers scattered in different journals. However, the book by Virchenko (1989), published in Russian, and made available to us by the author herself, covers some of the later works in a concise manner. This book does not appear to be wellknown to many researchers on dual integral equations although it is solely concerned with dual (triple) integral equations.

In this book, we present in a systematic manner some of the recent developments on dual integral equations involving various special functions as kernel. For dual integral equations, the kernels of both the integral equations involve the same special function and these usually occur while studying certain mixed boundary value problems involving homogeneous media in continuum mechanics. However, if one considers mixed boundary value problems involving piecewise non-homogeneous media, dual integral equations with principally different kernels may arise. These are termed *hybrid dual integral equations*. In this book, we also discuss some hybrid dual integral equations.

This book will be of interest to graduate students of applied mathematics, physics and engineering and to researchers in the theory of elasticity, fluid dynamics, mathematical physics, etc., who use dual integral equations as a mathematical tool.

This book is an outcome of research projects offered by the Council of Scientific and Industrial Research, New Delhi, to the authors during the last few years.

The authors thank Addison Wesley Longman for publishing this book in the Pitman Research Notes in Mathematics Series.

B. N. Mandal

Nanigopal Mandal

Chapter 1

Introduction

1.1   An overview of dual integral equations

Dual integral equations arise in the study of mixed boundary value problems in potential theory and the theory of diffusion in mathematical physics (cf. Sneddon 1966). The term *mixed boundary value problem* is used to distinguish this type of problem from the *uniform* problems of Dirichlet and Neumann. It may be mentioned that a problem in potential theory is said to be of Dirichlet type if the potential function inside a region Γ is prescribed at each point of $\partial\Gamma$, the boundary of Γ, and to be of Neumann type if the normal derivative of the potential function is prescribed on $\partial\Gamma$ but not the function itself. In potential theory, a typical problem of mixed type would be one in which the potential function is prescribed over a part of the boundary, and its normal derivative is prescribed over the remaining part. In another type of problem, the potential function is prescribed over a part of the boundary and a linear combination of the function and its normal derivative is prescribed over the remaining part.

To illustrate the formulation of a mixed boundary value problem which will result in dual integral equations, one can consider the simple problem of an electrified disc. The potential function is prescribed over the disc while outside the disc the surface charge density is zero. Use of the Hankel integral transform reduces the mixed boundary conditions to the following dual integral equations (cf. Sneddon 1966, p. 63):

$$\int_0^\infty f(t) \, J_0(xt) \, dt = u_0, \ 0 \le x < 1,$$

$$\int_0^\infty t \, f(t) \, J_0(xt) \, dt = 0, \ 1 < x < \infty,$$

where u_0 is the constant potential on the disc and $f(t)$ is an unknown function to be determined. The solution of this pair of integral equations was first obtained by Titchmarsh (1937) by applying Parseval's formula for the Mellin transform. Sneddon (1960) also obtained the solution of these dual integral equations by an elementary method.

The solution of the more general dual integral equations

$$\int_0^\infty t^\alpha f(t) \, J_\nu(xt) \ dt = g(x), \ 0 \le x < 1,$$

$$\int_0^\infty f(t) \, J_\nu(xt) \ dt = 0, \ 1 < x < \infty,$$

was first given by Titchmarsh (1937) for $\alpha > 0$ and by Busbridge (1938) for $\alpha > -2$ using Parseval's formula for the Mellin transform. Noble (1958, 1963) considered the dual integral equations of Titchmarsh type and obtained an explicit solution by applying the multiplying factor method based on Sonine's integrals of the first and second kind. These works are available in Sneddon's (1966) book. We present some later work on dual integral equations involving Bessel functions as kernels in Chapter 2 of this book.

Dual integral equations mentioned above usually arise when the mixed boundary value problems involve regions bounded by a circle or cylinder. However, if one wants to study a mixed boundary value problem involving other regions, say conical or spherical, then dual integral equations with Legendre or associated Legendre functions as kernel may occur. Babloian (1964) first considered some dual integral equations involving $P_{-1/2+i\tau}(\cosh x)$ as kernel. These are given below:

$$\int_0^\infty f(t)\, P_{-1/2+it}(\cosh x)\, dt = g(x),\ 0 \le x < a,$$

$$\int_0^\infty t^{\pm k} \tanh \pi t\, f(t)\, P_{-1/2+it}(\cosh x)\, dt = h(x),\ a < x < \infty,\ (k = \pm 1);$$

$$\int_0^\infty \{1 + \omega(t)\}\, f(t)\, P_{-1/2+it}(\cosh x)\, dt = g(x),\ 0 \le x < a,$$

$$\int_0^\infty t\, \tanh \pi t\, f(t)\, P_{-1/2+it}(\cosh x)\, dt = 0,\ a < x < \infty,$$

and

$$\int_0^\infty t\, f(t)\, P_{-1/2+it}(\cosh x)\, dt = 0,\ 0 \le x < a,$$

$$\int_0^\infty \{1 + \omega(t)\} \tanh \pi t\, f(t)\, P_{-1/2+it}(\cosh x)\, dt = h(x),\ a < x < \infty,$$

where $\omega(t)$ is a known weight function. He solved these pairs of integral equations by using integral representations involving $P_{-1/2+it}(\cosh x)$ together with the Mehler−Fock inversion theorem and obtained explicit solution to the first pair. For other pairs, these are reduced to solving appropriate Fredholm integral equations of the second kind. He also applied these dual integral equations to solve a torsion problem involving a region with spherical segment when the torsion is caused by rotating a small circular disc fastened rigidly at the centre of the flat part of the boundary plane. The details of these works are available in the treatise of Sneddon (1972). Later, various researchers contributed significantly to the theory and methods of solution of this class of dual integral equations involving Legendre or associated Legendre functions as kernels. In this book, we consider some of these later works systematically in Chapter 3.

Dual integral equations with trigonometic functions as kernel arise frequently in the analysis of mixed boundary value problems in a plane, such as crack problems in the two-dimensional theory of elasticity (cf. Sneddon and Lowengrub 1969). In Chapter 4, we discuss some recent developments on the theory and method of solution of a number of dual integral equations of this type.

Use of the Mehler−Fock integral transform to solve some classes of dual integral equations involving trigonometric functions as kernel was first made by Babloian

(1964). He obtained explicit solutions of the following pairs of integral equations by utilizing the Mehler–Fock inversion theorem:

$$\int_0^\infty t^k f(t) \sin xt \ dt = g(x), \ 0 \le x < a,$$

$$\int_0^\infty \coth \pi t \ f(t) \sin xt \ dt = h(x), \ a < x < \infty, \ (k = \pm 1),$$

and

$$\int_0^\infty f(t) \cos xt \ dt = g_1(x), \ 0 \le x < a,$$

$$\int_0^\infty t^k \coth \pi t \ f(t) \cos xt \ dt = h_1(x), \ a < x < \infty, \ (k = \pm 1).$$

These are discussed in the book by Sneddon (1972). In Chapter 4, we present some dual integral equations with trigonometric functions as kernel by applying the generalized Mehler–Fock inversion theorem (cf. Mandal and Mandal 1997, p. 47) and the generalized associated Mehler–Fock inversion theorem (cf. Mandal and Mandal 1997, p. 57) together with certain properties of associated and generalized associated Legendre functions. Solutions of the dual integral equations considered by Babloian (1964) are obtained as special cases.

Chapter 5 is devoted to some dual integral equations involving inverse Mellin transforms which arise in the solutions of certain mixed boundary value problems of potential theory for wedge-shaped regions.

All the dual integral equations mentioned above usually arise in the study of appropriate mixed boundary value problems involving a homogeneous medium in continuum mechanics. However, if one considers mixed boundary value problems involving piecewise non-homogeneous media, dual integral equations with principally different kernels, called *hybrid dual integral equations*, may arise. Virchenko (1984) considered some hybrid dual integral equations in which one equation contains the generalized associated Legendre function of the first kind $P_{-\frac{1}{2}+it}^{\mu,\nu}(\cosh x)$ as its kernel while the other equation contains a trigonometric function. This work is discussed in detail in Chapter 6.

In the book *Dual (Triple) Integral Equations* by N.A. Virchenko (1989), published in Russian, some of the dual integral equations, presented in this book, have also been discussed. However, this book appears to be not very wellknown to researchers in this area.

The mathematical analysis presented in this book is mostly formal and no attempt has been made to determine the precise conditions under which the solutions of the dual integral equations are valid. Whenever a change in the order of integration in a double integral is made, it is assumed that the concerned integral is convergent and the unknown function involved in it satisfies the necessary conditions which would make the integral convergent.

1.2 Two special methods for solving some classes of dual integral equations

In this section, we present two special methods for solving some classes of dual integral equations.

1.2.1 Method I

Mixed boundary value problems of the theory of elasticity, hydromechanics, aeromechanics, etc., and of mathematical physics for regions with partly infinite boundaries (namely, a strip, layer, cylinder, wedge, cone, etc.) can often be reduced to solving dual integral equations. These equations involve the use of integral transforms generated by the Sturm−Liouville problem on a semi-infinite interval. Here, we present a method to reduce a wide class of such dual integral equations to infinite algebraic systems of some special type. This method is due to Aleksandrov and Chebakov (1973).

Let us consider a second-order linear differential equation of the form:

$$(L - u^2)y = 0, \ 0 \le x < \infty, \tag{1.1}$$

5

where L is a Sturm–Liouville operator. Solving the corresponding Sturm–Liouville problem for this equation for $x \in [0, \infty)$, we construct the following integral transform:

$$g(x) \;=\; \int_0^\infty B(u, x)\, G(u)\, du, \tag{1.2}$$

$$G(u) \;=\; \int_0^\infty \overline{B}(u, s)\, g(s)\, ds, \tag{1.3}$$

where $B(u, x)$ is the eigenfunction of (1.1) for any $u \in [0, \infty)$, vanishing at infinity and bounded at zero.

Now we consider the dual integral equations

$$\int_0^\infty \omega(u)\, B(u, x)\, f(u)\, du \;=\; f_1(x), \; a \le x \le b, \tag{1.4}$$

$$\int_0^\infty B(u, x)\, f(u)\, du \;=\; 0, \; 0 < x < a, \; b < x < \infty, \tag{1.5}$$

where

$$\omega(u) = A\, \frac{F_1(u^2)}{F_2(u^2)} = A \prod_{n=1}^\infty (1 + \frac{u^2}{\delta_n^2})(1 + \frac{u^2}{\gamma_n^2})^{-1}, \; A = \text{ constant.} \tag{1.6}$$

The above form for $\omega(u)$ is an even meromorphic function where $i\delta_n$ and $i\gamma_n$ represent a denumerable set of zeros and poles of $\omega(u)$. We assume that there are no multiple zeros and poles and that $\delta_n \ne \gamma_m$ $(n, m = 1, 2, 3, \ldots)$. Also let δ_n and γ_n increase monotonically in absolute value with increasing n ensuring the convergence of the infinite product (1.6).

Let

$$\omega(u) = O(|u|^p), \; p \le 1. \tag{1.7}$$

If we apply the operator L to the function $B(u, x)$, then we obtain $u^2 B(u, x)$. Using this result and the expression for $\omega(u)$ from (1.6), the equation (1.4) can be written as

$$A\, F_1(L)\, F(x) = F_2(L)\, f_1(x), \; a \le x \le b, \tag{1.8}$$

where

$$F(x) = \int_0^\infty B(u, x) \, f(u) \, du, \qquad (1.9)$$

$F_1(L)$ and $F_2(L)$ being infinite order differential operators in x.

Solution of equation (1.8) can be expressed in the form

$$F(x) = \frac{F_2(L)}{AF_1(L)} \, f_1(x) + \sum_{n=1}^\infty [C_n \, B(i\delta_n, x) + D_n \, E(i\delta_n, x)], \ a \le x \le b. \qquad (1.10)$$

The first term in (1.10) represents a particular solution of the inhomogeneous equation (1.8) and the infinite sum gives the general solution of the homogeneous equation. The functions $B(u, x)$ and $E(u, x)$ are two linearly independent solutions of (1.1). From the relations (1.5), (1.9) and (1.10), we have

$$f(u) = \int_a^b F(s) \, \overline{B}(u, s) \, ds. \qquad (1.11)$$

Assuming that the function $f_1(x)$ can, in general, be represented by the integral (1.2) or approximated by means of a linear combination of the functions $B(\epsilon_k, x)(k = 1, 2, 3, \ldots, N)$, we set

$$f_1(x) = B(\epsilon, x).$$

The first term in (1.10) now assumes the form $\omega^{-1}(\epsilon)B(\epsilon, x)$. Then substituting the expression for $F(x)$ from (1.10) into (1.11), we find that

$$f(u) = \omega^{-1}(\epsilon) \, \phi(u, -i\epsilon) + \sum_{n=1}^\infty [C_n \, \phi(u, \delta_n) + D_n \, \psi(u, \delta_n)], \qquad (1.12)$$

where

$$\phi(u, x) = \int_a^b B(ix, s) \, \overline{B}(u, s) \, ds, \ \psi(u, x) = \int_a^b E(ix, s) \, \overline{B}(u, s) \, ds.$$

The unknown constants C_n and D_n are to be determined from the condition that the solution (1.12) satisfies the dual integral equations (1.4) and (1.5).

We may note here that under the assumptions made above, the meromorphic function $\omega(u)$ can be represented by

$$\omega(u) = A - \frac{2}{\pi} \sum_{k=1}^\infty \frac{u^2 \, s_k}{\gamma_k(u^2 + \gamma_k^2)}, \qquad (1.13)$$

7

where

$$s_k = \frac{\pi i}{[\omega^{-1}(i\gamma_k)]'}, \ A = \frac{2}{\pi} \sum_{k=1}^{\infty} \frac{s_k}{\gamma_k} \ \text{for } p < 0.$$

Substituting (1.12) into (1.4) and using the relations

$$\int_0^{\infty} \phi(u, -i\epsilon) \, B(u, x) \, du = f_1(x) = B(\epsilon, x), \ a \leq x \leq b,$$

and $\omega(i\delta_n) = 0$, we obtain an infinite linear algebraic system for the unknowns C_n and D_n as

$$\omega^{-1}(\epsilon) \int_0^{\infty} \phi(u, -i\epsilon) \, \{\omega(u) - \omega(\epsilon)\} \, B(u, x) \, du$$

$$+ \sum_{n=1}^{\infty} \left\{ C_n \int_0^{\infty} \phi(u, \delta_n) \, [\omega(u) - \omega(i\delta_n)] \, B(u, x) \, du \right.$$

$$\left. + D_n \int_0^{\infty} \psi(u, \delta_n) \, [\omega(u) - \omega(i\delta_n)] \, B(u, x) \, du \right\} = 0.$$

Using (1.13), the above system becomes

$$\sum_{k=1}^{\infty} s_k \, \gamma_k \left\{ \frac{\omega^{-1}(\epsilon)}{\gamma_k^2 + \epsilon^2} \, T_k(x, -i\epsilon) + \sum_{n=1}^{\infty} \frac{1}{\gamma_k^2 - \delta_n^2} \right.$$

$$\left. \times \, [C_n \, T_k(x, \delta_n) + D_n \, U_k(x, \delta_n)] \right\} = 0, \ a \leq x \leq b, \tag{1.14}$$

where

$$T_k(x, \nu) = \int_0^{\infty} \frac{u^2 + \nu^2}{u^2 + \gamma_k^2} \, \phi(u, \nu) \, B(u, x) \, du,$$

$$U_k(x, \nu) = \int_0^{\infty} \frac{u^2 + \nu^2}{u^2 + \gamma_k^2} \, \psi(u, \nu) \, B(u, x) \, du.$$

Now the solution of (1.8) with respect to the known function $f_1(x)$ can be written in the form

$$f_1(x) = \frac{A F_1(L)}{F_2(L)} \, F(x) + \sum_{k=1}^{\infty} [G_k \, B(i\gamma_k, x) + H_k \, E(i\gamma_k, x)], \ a \leq x \leq b. \tag{1.15}$$

The constants G_k and H_k can be obtained from the condition that (1.15) is equivalent to the equation (1.4) and hence to (1.14). G_k and H_k are linear functionals in $F(x)$.

Now putting the expression for $F(x)$ from (1.10) into (1.15) and using the conditions that

$$\frac{AF_1(L)}{F_2(L)}\ B(i\delta_n, x)\ =\ \omega(i\delta_n)\ B(i\delta_n, x) = 0,$$

$$\frac{AF_1(L)}{F_2(L)}\ E(i\delta_n, x)\ =\ \omega(i\delta_n)\ E(i\delta_n, x) = 0,$$

the equation (1.15) and hence (1.14) becomes

$$\sum_{k=1}^{\infty}\sum_{n=1}^{\infty}\{(C_n\ G_{kn}^+ + D_n\ H_{kn}^+)B(i\gamma_k, x) + (C_n\ G_{kn}^- + D_n\ H_{kn}^-)E(i\gamma_k, x)\}$$

$$+\omega^{-1}(\epsilon)\sum_{k=1}^{\infty}\{G_k^*\ B(i\gamma_k, x) + H_k^*\ E(i\gamma_k, x)\} = 0,$$

where

$$G_k^*\ =\ G_k\ B(\epsilon, x),\ H_k^* = H_k\ B(\epsilon, x),$$
$$G_{kn}^+\ =\ G_k\ B(i\delta_n, x),\ H_{kn}^+ = H_k\ B(i\delta_n, x),$$
$$G_{kn}^-\ =\ G_k\ E(i\delta_n, x),\ H_{kn}^- = H_k\ E(i\delta_n, x).$$

Finally assuming that the functions $B(i\gamma_k, x)$ and $E(i\gamma_k, x)$ $(k = 1, 2, 3, \ldots)$ are linearly independent, we obtain the following two infinite algebraic systems for the unknown constants appearing in the relation (1.10) of the dual integral equations as given by

$$\omega^{-1}(\epsilon)\ G_k^* + \sum_{n=1}^{\infty}(C_n\ G_{kn}^+ + D_n\ H_{kn}^+)\ =\ 0, \tag{1.16}$$

$$\omega^{-1}(\epsilon)\ H_k^* + \sum_{n=1}^{\infty}(C_n\ G_{kn}^- + D_n\ H_{kn}^-)\ =\ 0, \tag{1.17}$$

where $k = 1, 2, 3, \ldots$

The above method is now illustrated for two forms of the operator L.

(i) Let the operator L be of the form

$$L = x^2 - x\ \frac{d}{dx} - x^2\ \frac{d^2}{dx^2}, \ 0 < x < \infty. \tag{1.18}$$

Then the functions $B(u, x)$ and $\overline{B}(u, s)$ in the integral transform formulae (1.2) and (1.3) have the form

$$B(u, x) = K_{iu}(x)\ \text{ and }\ \overline{B}(u, s) = \frac{2u \sinh \pi u}{\pi^2 s}\ K_{iu}(s), \tag{1.19}$$

9

where $K_{iu}(x)$ is Macdonald's function. The transform obtained in this case is the Kontorovich–Lebedev integral transform. The dual integral equations (1.4) and (1.5) generated by this integral transform can be reduced to (1.14) in which

$$\phi(u, x) = \frac{2u \sinh \pi u}{\pi^2} \int_a^b K_{-x}(s) \, K_{iu}(s) \, \frac{ds}{s}, \tag{1.20}$$

$$\psi(u, x) = \frac{2u \sinh \pi u}{\pi^2} \int_a^b I_{-x}(s) \, K_{iu}(s) \, \frac{ds}{s}, \tag{1.21}$$

where $I_{-x}(s) = E(ix, s)$ is the Bessel function of the second kind, and the solution of this pair can be obtained in the form of (1.12). Now the relation (1.14) can be reduced to an infinite system of linear algebraic equations with respect to the unknown constants C_n and D_n given in equation (1.10), by computing the integrals in (1.20), (1.21) and (1.14). The integrals in (1.20) and (1.21) can be found by using the well-known relations given by Watson (1944), while the computation of the integrals in (1.14) can be reduced to obtaining an integral of the form

$$R(x, c) = \int_0^\infty \frac{\tau \sinh \pi \tau}{\tau^2 + \gamma_k^2} \, K_{i\tau}(x) \, K_{i\tau}(c) \, d\tau. \tag{1.22}$$

Using the integral representations

$$K_{i\tau}(x) = \cosh^{-1}\left(\frac{\pi \tau}{2}\right) \int_0^\infty \cos(x \cosh s) \cos \tau s \, ds,$$

$$K_{i\tau}(x) = \cosh^{-1}\left(\frac{\pi \tau}{2}\right) \int_0^\infty \sin(x \sinh t) \sin \tau t \, dt,$$

and

$$I_{i\tau}(x) = \frac{2}{\pi} \cosh\left(\frac{\pi \tau}{2}\right) \int_0^\infty \sin(x \sinh t) \, e^{-i\tau t} \, dt,$$

we obtain from (1.22), for any γ_k

$$R(x, c) = \begin{cases} \dfrac{\pi^2}{2} \, I_{\gamma_k}(x) \, K_{\gamma_k}(c), & 0 < x < a, \\[4mm] \dfrac{\pi^2}{2} \, K_{\gamma_k}(x) \, I_{\gamma_k}(c), & c < x < \infty. \end{cases}$$

Substituting the values of the functions $T_k(x, \nu)$ and $U_k(x, \nu)$ thus obtained into the relation (1.14) and equating to zero the coefficients of the linearly independent functions $K_{\gamma_k}(x)$ and $I_{\gamma_k}(x)$, we obtain the following infinite systems for the unknown constants C_n and D_n:

$$\frac{\omega^{-1}(\epsilon)}{\epsilon^2 + \gamma_k^2} \left[K'_{i\epsilon}(b) K_{\gamma_k}(b) - K_{i\epsilon}(b) K'_{\gamma_k}(b) \right]$$

$$+ \sum_{n=1}^{\infty} \frac{C_n}{\delta_n^2 - \gamma_k^2} \left[K'_{-\delta_n}(b) K_{\gamma_k}(b) - K_{-\delta_n}(b) K'_{\gamma_k}(b) \right]$$

$$+ \sum_{n=1}^{\infty} \frac{D_n}{\delta_n^2 - \gamma_k^2} \left[I'_{-\delta_n}(b) K_{\gamma_k}(b) - I_{-\delta_n}(b) K'_{\gamma_k}(b) \right] = 0,$$

and

$$\frac{\omega^{-1}(\epsilon)}{\epsilon^2 + \gamma_k^2} \left[K'_{i\epsilon}(a) I_{\gamma_k}(a) - K_{i\epsilon}(a) I'_{\gamma_k}(a) \right]$$

$$+ \sum_{n=1}^{\infty} \frac{C_n}{\delta_n^2 - \gamma_k^2} \left[K'_{-\delta_n}(a) I_{\gamma_k}(a) - K_{-\delta_n}(a) I'_{\gamma_k}(a) \right]$$

$$+ \sum_{n=1}^{\infty} \frac{D_n}{\delta_n^2 - \gamma_k^2} \left[I'_{-\delta_n}(a) I_{\gamma_k}(a) - I_{-\delta_n}(a) I'_{\gamma_k}(a) \right] = 0,$$

where $k = 1, 2, 3, \ldots$

(ii) Let us assume

$$L = (1 - t^2) \frac{d^2}{dt^2} - 2t \frac{d}{dt} - \frac{m^2}{1 - t^2} - \frac{1}{4}, \ t = \cosh x, \ 0 \le x < \infty,$$

where m is a positive integer or zero. Then the functions $B(u, x)$ and $\overline{B}(u, s)$ in the transform formulae (1.2) and (1.3) become

$$B(u, x) = P^m_{-1/2+iu}(\cosh x), \ \overline{B}(u, s) = u \tanh \pi u \, P^{-m}_{-1/2+iu}(\cosh s), \qquad (1.23)$$

where $P^m_{-1/2+iu}(\cosh x)$ is the associated Legendre function of the first kind. In this case, the transform obtained is the Mehler–Fock integral transform. The dual integral equations (1.4) and (1.5) generated by this transform for $a = 0$ become

11

$$\int_0^\infty \omega(\tau)\, P_{-1/2+i\tau}^m(\cosh x)\, f(\tau)\, d\tau \;=\; f_1(x),\; 0 \le x \le b, \qquad (1.24)$$

$$\int_0^\infty P_{-1/2+i\tau}^m(\cosh x)\, f(\tau)\, d\tau \;=\; 0,\; b < x < \infty. \qquad (1.25)$$

By the above method, the solution of this pair can be represented by

$$f(\tau) = \tau \tanh \pi\tau \int_0^b F(t)\, P_{-1/2+i\tau}^{-m}(\cosh t)\, \sinh t\, dt, \qquad (1.26)$$

where

$$F(x) = \frac{F_2(L)}{AF_1(L)} f_1(x) + \sum_{n=1}^\infty \left[C_n P_{-1/2-\delta_n}^m(\cosh x) + D_n Q_{-1/2-\delta_n}^m(\cosh x) \right],\; 0 \le x \le b.$$

Now since $F(x)$ is bounded at $x = 0$ and $Q_{-1/2-\delta_n}^m(\cosh x)$ has a logarithmic singularity at $x = 0$, we choose $D_n = 0$ $(n = 1, 2, 3, \ldots)$. We also assume that $m = 0$, since when $m \ne 0$, the dual integral equations (1.24) and (1.25) can be reduced to the pair (1.24) and (1.25) with $m = 0$ by using the relation

$$P_{-1/2+i\tau}^m(x) = (x^2 - 1)^{\frac{m}{2}}\, \frac{d^m P_{-1/2+i\tau}(x)}{dx^m}. \qquad (1.27)$$

We now reduce the expression (1.14) to an infinite algebraic system for the unknown constants C_n from (1.26), by finding $\phi(u, x)$ from (1.12) and $T_k(x, \nu)$ from (1.14).

Here, we have

$$\phi(u, x) = u \tanh \pi u \int_0^b P_{-1/2-x}(\cosh t)\, P_{-1/2+iu}(\cosh t)\, \sinh t\, dt$$

$$= \frac{\sinh b}{x^2 + u^2} \left[P_{-1/2+iu}(\cosh b)\, P_{-1/2-x}^1(\cosh b) \right]$$

$$- P_{-1/2-x}(\cosh b)\, P_{-1/2+iu}^1(\cosh b) \right] u \tanh \pi u. \qquad (1.28)$$

Using (1.28), we can reduce the problem of finding $T_k(x, \nu)$ to that of computing the integrals of the form

$$R_1(x, b) = \int_0^\infty \frac{\tau \tanh \pi\tau}{\tau^2 + \gamma_k^2}\, P_{-1/2+i\tau}(\cosh x)\, P_{-1/2+i\tau}(\cosh b)\, d\tau.$$

Utilizing the integral representations of the functions $P_{-1/2+i\tau}(\cosh x)$ and $Q_{-1/2+i\tau}(\cosh b)$ (cf. Hobson 1931), we have

$$R_1(x,b) = \begin{cases} P_{-\frac{1}{2}-\gamma_k}(\cosh x) \, Q_{-\frac{1}{2}+\gamma_k}(\cosh b), & x < b, \\ Q_{-\frac{1}{2}+\gamma_k}(\cosh x) \, P_{-\frac{1}{2}-\gamma_k}(\cosh b), & b < x. \end{cases}$$

Substituting the function $T_k(x,\nu)$ obtained in this way into the equation (1.14) and equating to zero the coefficients of the linearly independent functions $P_{-\frac{1}{2}-\gamma_k}(\cosh x)$ $(k = 1, 2, 3, \ldots)$, we get the following infinite system of equations for the unknown constants C_n :

$$\frac{\omega^{-1}(\epsilon)}{\epsilon^2 + \gamma_k^2} \left\{ P_{-\frac{1}{2}+i\epsilon}(\cosh b) Q^1_{-\frac{1}{2}+\gamma_k}(\cosh b) - P^1_{-\frac{1}{2}+i\epsilon}(\cosh b) Q_{-\frac{1}{2}+\gamma_k}(\cosh b) \right\}$$

$$+ \sum_{n=1}^{\infty} \frac{C_n}{\delta_n^2 - \gamma_k^2} \left\{ P^1_{-\frac{1}{2}-\delta_n}(\cosh b) Q_{-\frac{1}{2}+\gamma_k}(\cosh b) - P_{-\frac{1}{2}-\delta_n}(\cosh b) Q^1_{-\frac{1}{2}+\gamma_k}(\cosh b) \right\} = 0,$$

where $k = 1, 2, 3, \ldots$.

Aleksandrov (1975) considered a simple generalization of the differential operator L and followed the same method presented above to reduce the dual integral equations

$$\int_{-\infty}^{\infty} \omega(u) \, f(u) \, B(u,x) \, d\rho(u) \;=\; f_1(x), \; c \leq x \leq d,$$

$$\int_{-\infty}^{\infty} f(u) \, B(u,x) \, d\rho(u) \;=\; 0, \; a \leq x \leq c, \; d < x < \infty,$$

where $\omega(u)$ is defined by (1.6) and $\rho(u)$ is nondecreasing spectrum function of L, to an infinite system of linear algebraic equations.

1.2.2 Method II

Dual integral equations involving various types of special functions as kernel can sometimes be solved by using a simple method. By defining an appropriate integral transform of a function related to the unknown function satisfying the dual integral equations, it is observed that the transformed function satisfies a non-homogeneous ordinary differential equation whose solution is easy to find in many cases. Then the inverse transformation produces ultimately the function sought for. Chakrabarti (1989) used this method to solve some dual integral equations involving the first-kind Bessel function of order one as kernel. These dual integral equations are considered in §2.1.3. Srivastava (1990) used this technique for some dual integral equations also involving the first-kind Bessel function as kernel. The related integral transform used here is the Hankel transform. Mandal and Mandal (1993) used this technique for some classes of dual integral equations with Legendre and associated Legendre functions as kernel. The related integral transforms utilized are the ordinary and the generalized Mehler−Fock integral transform. These are discussed in detail in Chapter 3 of this book.

As an illustration of this method, we consider here the following dual integral equations which are wellstudied in the literature (cf. Sneddon 1966, p. 77; Davis 1991):

$$\int_0^\infty t\, f(t)\, J_0(xt)\, dt \;=\; \pi/2,\; 0 \le x \le 1, \tag{1.29}$$

$$\int_0^\infty f(t)\, J_0(xt)\, dt \;=\; 0,\; 1 < x < \infty, \tag{1.30}$$

where $f(t)$ is an unknown function to be determined. Putting $f(t) = t\, \psi(t)$, the above pair reduces to the form

$$\int_0^\infty t^2\, \psi(t)\, J_0(xt)\, dt \;=\; \pi/2,\; 0 \le x \le 1, \tag{1.31}$$

$$\int_0^\infty t\, \psi(t)\, J_0(xt)\, dt \;=\; 0,\; 1 < x < \infty. \tag{1.32}$$

Now we set

$$\int_0^\infty t\, \psi(t)\, J_0(xt)\, dt = \phi(x),\; 0 \le x \le 1, \tag{1.33}$$

where $\phi(x)$ is an unknown auxiliary function. Then, by the Hankel inversion theorem, equations (1.32) and (1.33) give

$$\psi(t) = \int_0^1 \xi\, \phi(\xi)\, J_0(\xi t)\, d\xi. \tag{1.34}$$

14

Substituting the above expression for $\psi(t)$ into equation (1.31), we get

$$\int_0^\infty t^2 \, J_0(xt) \left\{ \int_0^1 \xi \, \phi(\xi) \, J_0(\xi t) \, d\xi \right\} \, dt = \pi/2, \; 0 \le x \le 1. \tag{1.35}$$

Since $u = J_0(xt)$ satisfies the ordinary differential equation (ODE)

$$\frac{d^2u}{dx^2} + \frac{1}{x}\frac{du}{dx} + t^2 u = 0,$$

the equation (1.35) is equivalent to the ODE

$$\frac{d^2v}{dx^2} + \frac{1}{x}\frac{dv}{dx} = -\pi/2, \; 0 \le x \le 1, \tag{1.36}$$

where

$$v(x) = \int_0^\infty J_0(xt) \left\{ \int_0^1 \xi \, \phi(\xi) \, J_0(\xi t) \, d\xi \right\} dt. \tag{1.37}$$

Solution of the ODE (1.36), by the method of variation of parameters, is

$$v(x) = A + B \log x - \frac{\pi x^2}{8}, \; 0 \le x \le 1, \tag{1.38}$$

where A and B are arbitrary constants to be determined. The constant B must be equal to zero since $v(x)$ is to be finite at $x = 0$. The other constant A will be determined later. Hence,

$$v(x) = A - \frac{\pi x^2}{8}, \; 0 \le x \le 1. \tag{1.39}$$

Equation (1.37) can be written as

$$v(x) = \int_0^1 \xi \, \phi(\xi) \left\{ \int_0^\infty J_0(xt) \, J_0(\xi t) \, dt \right\} d\xi, \; 0 \le x \le 1. \tag{1.40}$$

Using the result (A.1.15) given in the Appendix and then interchanging the order of integration, we get an Abel-type integral equation

$$\frac{2}{\pi} \int_0^x \frac{ds}{\sqrt{x^2 - s^2}} \left\{ \int_s^1 \frac{\xi \, \phi(\xi)}{\sqrt{\xi^2 - s^2}} \, d\xi \right\} = v(x), \; 0 \le x \le 1. \tag{1.41}$$

By repeated application of Abel inversion theorems together with the use of the relation (1.39), we obtain the unknown function $\phi(x)$ from the equation (1.41) as given by

$$\phi(x) = \frac{4A - \pi}{2\pi\sqrt{1 - x^2}} + \sqrt{1 - x^2}, \; 0 \le x \le 1. \tag{1.42}$$

15

The unknown constant A is now determined from the fact that $\phi(x)$ is continuous at $x = 1$ so that $A = \pi/4$. Therefore, for such a value of A, $\phi(x)$ becomes

$$\phi(x) = \sqrt{1 - x^2}, \ 0 \le x \le 1. \tag{1.43}$$

Hence, the solution of the dual integral equations (1.31) and (1.32) is obtained by the relations (1.34) and (1.43) as

$$\psi(t) = \int_0^\infty \xi \sqrt{1 - \xi^2} \, J_0(\xi t) \, d\xi$$

$$= \frac{1}{t^2} \left(\frac{\sin t}{t} - \cos t \right).$$

Thus, the solution of the pair (1.29) and (1.30) is given by

$$f(t) = \frac{\sin t}{t^2} - \frac{\cos t}{t}.$$

The above solution completely agrees with Davis (1991) and Sneddon (1966).

Chapter 2

Dual integral equations with Bessel function kernel

Dual integral equations involving Bessel functions as kernel have been studied systematically in the treatise by Sneddon (1966). In this chapter, we present methods of solution of some dual integral equations with Bessel function kernel which are not available in any book or monograph.

2.1 Kernels involving a Bessel function of the first kind

In this section, dual integral equations with a Bessel function of the first kind as kernel are presented. First we consider some dual integral equations involving a Bessel function of the first kind of order zero. The solutions of these pairs of integral equations are obtained by exploiting some properties of the Bessel functions.

2.1.1 Nasim and Aggarwala (1984) considered the following simple pair of dual integral equations of Beltrami type:

$$\int_0^\infty t^{-1} f(t)\; J_0(xt)\; dt \;=\; g(x),\; 0 < x < 1, \tag{2.1}$$

$$\int_0^\infty f(t)\; J_0(xt)\; dt \;=\; 0,\; 1 < x < \infty. \tag{2.2}$$

This pair of integral equations arise in the problem of the electrified disc mentioned in §1.1. To obtain the solution of this pair, we express (2.2) as

$$\int_0^\infty t^{-\frac{1}{2}} f(t)\, J_0(xt)\, (xt)^{\frac{1}{2}}\, dt = \phi(x)\, H(1-x),\ 0 < x < \infty, \tag{2.3}$$

where $\phi(x)$ is an unknown function to be determined, $H(x)$ being the Heaviside unit function. Then, by the Hankel inversion theorem, we find that

$$t^{-\frac{1}{2}} f(t) = \int_0^1 \phi(x)\, J_0(xt)\, (xt)^{\frac{1}{2}}\, dx. \tag{2.4}$$

Using the representation (A.1.8) for $J_0(xt)$, this equation can be written as

$$t^{-\frac{1}{2}} f(t) = \frac{2}{\pi} \int_0^1 \phi(x)(xt)^{\frac{1}{2}} \left\{ \int_0^x \frac{\cos ut}{\sqrt{x^2 - u^2}}\, du \right\} dx$$

$$= \frac{2\sqrt{t}}{\pi} \int_0^1 \cos ut \left\{ \int_u^1 \frac{\sqrt{x}\, \phi(x)}{\sqrt{x^2 - u^2}}\, dx \right\} du.$$

Therefore,

$$f(t) = t \int_0^1 \psi(u) \cos ut\, du, \tag{2.5}$$

where

$$\psi(u) = \frac{2}{\pi} \int_u^1 \frac{\sqrt{x}\, \phi(x)}{\sqrt{x^2 - u^2}}\, dx. \tag{2.6}$$

Substituting the expression for $f(t)$ from (2.5) into (2.1), we obtain

$$g(x) = \int_0^\infty J_0(xt) \left\{ \int_0^1 \psi(u) \cos ut\, du \right\} dt$$

$$= \int_0^1 \psi(u) \left\{ \int_0^\infty \cos ut\, J_0(xt)\, dt \right\} du,\ 0 < x < 1. \tag{2.7}$$

Using the result (A.1.1), the above equation reduces to an Abel integral equation

$$g(x) = \int_0^x \frac{\psi(u)}{\sqrt{x^2 - u^2}}\, du,\ 0 < x < 1. \tag{2.8}$$

The solution of this integral equation is

$$\psi(u) = \frac{2}{\pi} \frac{d}{du} \int_0^u \frac{x\, g(x)}{\sqrt{u^2 - x^2}}\, dx,\ 0 < u < 1. \tag{2.9}$$

18

Thus, from the relations (2.5) and (2.9), the solution of the dual integral equations (2.1) and (2.2) is obtained as

$$f(t) = \frac{2t}{\pi} \int_0^1 \cos ut \left\{ \frac{d}{du} \int_0^u \frac{x \, g(x) \, dx}{\sqrt{u^2 - x^2}} \right\} du. \tag{2.10}$$

A more familiar form of the above solution is obtained by integrating the u-integral by parts and then simplifying, and is given by

$$f(t) = \frac{2t}{\pi} \left[\cos t \int_0^1 \frac{x \, g(x)}{\sqrt{1 - x^2}} \, dx + t \int_0^1 \frac{s \, ds}{\sqrt{1 - s^2}} \left\{ \int_0^1 u \sin ut \, f(us) \, du \right\} \right]. \tag{2.11}$$

In particular, if $g(x) = 1$, the solution of the corresponding dual integral equations (2.1) and (2.2) becomes

$$f(t) = \frac{2}{\pi} \sin t.$$

2.1.2 Here we consider two pairs of dual integral equations with Bessel function of zeroth order as kernels and describe the methods of solution. These methods exploit the fact that, under certain circumstances of practical importance, one of the integrals of each pair of the dual integral equations under consideration possesses a *square-root* singularity at the *turning point*. By the term *turning point*, we understand the point where the boundary condition, in the mixed boundary value problem in which these dual integral equations arise, changes. This will become clear as we proceed further in our analysis. Utilization of such singular behaviour of one of the integrals of a pair of integral equations has been extensively used by Davis (1991) for a class of mixed boundary value problems arising in viscous fluid flow theory, and also by Chakrabarti and Mandal (1998) recently.

First, we consider the following dual integral equations:

$$\int_0^\infty t \, f(t) \, J_0(xt) \, dt = \pi/2, \ 0 < x < 1, \tag{2.12}$$

$$\int_0^\infty f(t) \, J_0(xt) \, dt = 0, \ 1 < x < \infty, \tag{2.13}$$

where $f(t)$ is the unknown function to be determined. To obtain the solution of this pair, we set

$$\int_0^\infty t \, f(t) \, J_0(xt) \, dt = \frac{\pi}{2x} \frac{d}{dx} \int_1^x \frac{s \, \psi(s)}{\sqrt{x^2 - s^2}} \, ds, \ 1 < x < \infty, \tag{2.14}$$

19

where $\psi(t)$ is another unknown differentiable function to be determined, such that $\psi(1) \neq 0$, and this exhibits appropriate singularity behaviour of the integral on the left at the turning point $x = 1$.

Utilizing the Hankel inversion theorem along with the results (A.1.4) and (A.1.7), from the relations (2.12) and (2.14), we find that

$$f(t) = \frac{\pi}{2}\left[\frac{J_1(t)}{t} + \int_1^\infty \psi(s)\sin ts\ ds\right]. \tag{2.15}$$

Substituting the expression for $f(t)$ from (2.15) into the equation (2.13) and then using the result (A.1.2), we obtain an Abel type integral equation as given by

$$\int_x^\infty \frac{\psi(s)}{\sqrt{s^2 - x^2}}\ ds = -\int_0^\infty \frac{J_0(xt)\ J_1(t)}{t}\ dt,\ 1 < x < \infty. \tag{2.16}$$

Using Abel's inversion formula and then interchanging the order of integration, the unknown function $\psi(s)$ is obtained as

$$\psi(s) = \frac{2}{\pi}\int_0^\infty \frac{J_1(t)}{t}\left\{\frac{d}{ds}\int_s^\infty \frac{x\ J_0(xt)}{\sqrt{x^2 - s^2}}\ dx\right\}\ dt,\ 1 < s < \infty.$$

Then, using the result (A.1.5), the above relation reduces to

$$\psi(s) = -\frac{2}{\pi(s + \sqrt{s^2 - 1})},\ 1 < s < \infty. \tag{2.17}$$

Substituting the above expression for $\psi(s)$ into equation (2.15) and simplifying with the use of the result (A.1.10), the solution of the dual integral equations (2.12) and (2.13) is obtained as given by

$$f(t) = \frac{\sin t}{t^2} - \frac{\cos t}{t} = -\frac{d}{dt}\left(\frac{\sin t}{t}\right). \tag{2.18}$$

The above result agrees with the one cited in Davis's paper (1991) as well as in Sneddon's book (1966), p. 77. This solution is also obtained by a special method mentioned in §1.2.2.

Next we consider the pair of integral equations

$$\int_0^\infty f(t)\ J_0(xt)\ dt = \pi/2,\ 0 < x < 1, \tag{2.19}$$

$$\int_0^\infty t\ f(t)\ J_0(xt)\ dt = 0,\ 1 < x < \infty. \tag{2.20}$$

20

For the solution of the above pair, we set

$$\int_0^\infty t\, f(t)\, J_0(xt)\, dt = \frac{1}{x}\frac{d}{dx}\int_x^1 \frac{s\,\psi(s)}{\sqrt{s^2-x^2}}\, ds,\ 0 < x < 1, \tag{2.21}$$

where now $\psi(s)$ is a new unknown differentiable function with $\psi(1)\neq 0$ and is to be determined.

By the Hankel inversion theorem and use of the result (A.1.6), from equations (2.20) and (2.21), the solution of the above dual integral equations is obtained, in terms of the unknown function $\psi(s)$, as

$$f(t) = -\int_0^1 \psi(s)\,\cos ts\, ds. \tag{2.22}$$

Using the above expression for $f(t)$ and the result (A.1.1), the equation (2.19) reduces to the following Abel type integral equation

$$\int_0^x \frac{\psi(s)}{\sqrt{x^2-s^2}}\, ds = -\frac{\pi}{2},\ 0 < x < 1, \tag{2.23}$$

the solution of which is given by

$$\psi(s) = -1,\ 0 < s < 1. \tag{2.24}$$

Thus from the relations (2.22) and (2.24), the solution of the pair of integral equations (2.19) and (2.20) is obtained in the form

$$f(t) = \frac{\sin t}{t}. \tag{2.25}$$

The above result agrees with the corresponding result available in Sneddon's book (1966), p. 64.

2.1.3 Here, we consider some dual integral equations involving Bessel functions of the first kind of order one which are of physical interest. Chakrabarti (1989) considered these dual integral equations and solved them by the method mentioned in §1.2.2. We consider the following pairs of integral equations.

(i) First we consider the dual integral equations

$$\int_0^\infty t^2\, f(t)\, J_1(xt)\, dt \;=\; -g(x),\; 0 < x < a, \tag{2.26}$$

$$\int_0^\infty t\, f(t)\, J_1(xt)\, dt \;=\; 0,\; x > a. \tag{2.27}$$

This pair arises in the study of a static penny-shaped crack problem in a homogeneous isotropic elastic solid under torsion. $g(x)$ represents the distribution of shear stress on the face of the crack and it is required that the displacement field given by the integral on the left of (2.27) is zero at $x = a$ for continuity requirements.

To obtain the solution of this pair of integral equations, we set

$$\int_0^\infty t\, f(t)\, J_1(xt)\, dt = h(x),\; 0 < x < a. \tag{2.28}$$

Thus, by the Hankel inversion theorem, the relations (2.27) and (2.28) produce

$$f(t) = \int_0^a s\, h(s)\, J_1(st)\, ds. \tag{2.29}$$

Substituting the expression for $f(t)$ from the relation (2.29) into equation (2.26), then interchanging the order of integration, we get

$$\int_0^\infty t^2 J_1(xt) \left\{ \int_0^a s\, h(s)\, J_1(st)\, ds \right\} dt = -g(x),\; 0 < x < a. \tag{2.30}$$

The above integral equation (2.30) is equivalent to the ODE

$$x^2\, \frac{d^2 u}{dx^2} + x\, \frac{du}{dx} - u(x) = x^2\, g(x),\; 0 < x < a, \tag{2.31}$$

where

$$u(x) = \int_0^\infty J_1(xt) \left\{ \int_0^a s\, h(s)\, J_1(st)\, ds \right\} dt. \tag{2.32}$$

By the method of variation of parameters, the general solution of the ODE (2.31) can be obtained as given by

$$u(x) = A\, x + \frac{B}{x} + \frac{x}{2} \int_0^x g(t)\, dt - \frac{1}{2x} \int_0^x t^2\, g(t)\, dt, \tag{2.33}$$

where A and B are arbitrary constants to be determined from the physical considerations of the problem. The constant B must be zero as $u(x)$ is finite at $x = 0$, and the constant A will be determined later.

Now interchanging the order of integration and then using the result (A.1.15), the relation (2.32) reduces to an Abel integral equation of the first kind as given by

$$\frac{2}{\pi} \int_0^x \frac{t^2}{\sqrt{x^2 - t^2}} \left\{ \int_t^a \frac{h(s)\, ds}{\sqrt{s^2 - t^2}} \right\} dt = x\, u(x), \ 0 < x < a. \tag{2.34}$$

By Abel's inversion theorem, we obtain another Abel integral equation of the second kind as

$$t \int_t^a \frac{h(s)\, ds}{\sqrt{s^2 - t^2}} = \int_0^t \frac{\frac{d}{dx}\{x\, u(x)\}}{\sqrt{t^2 - x^2}}\, dx. \tag{2.35}$$

Again applying Abel's inversion theorem, equation (2.35) gives

$$h(s) = \frac{2s}{\pi a \sqrt{a^2 - s^2}} \int_0^a \frac{\frac{d}{dx}\{x\, u(x)\}}{\sqrt{a^2 - x^2}}\, dx - \frac{2s}{\pi} \int_s^a \frac{\frac{d}{dx}\{v(x)\}}{\sqrt{x^2 - s^2}}\, dx, \tag{2.36}$$

where

$$v(x) = \frac{1}{x} \int_0^a \frac{\frac{d}{dy}\{y\, u(y)\}}{\sqrt{x^2 - y^2}}\, dy. \tag{2.37}$$

Therefore, equations (2.36) and (2.37) along with equation (2.33) solve the integral equation (2.30) completely if the constant A is determined. The constant A is determined from the fact that $h(a) = 0$ for its continuity at $x = a$. Therefore, the constant A will be determined by the relation

$$\int_0^a \frac{\frac{d}{dx}\{x\, u(x)\}}{\sqrt{a^2 - x^2}}\, dx = 0. \tag{2.38}$$

As a particular case, taking $g(x) = x$, we find that

$$u(x) = A + \frac{x^2}{8}.$$

Then the relation (2.38) gives

$$A = -\frac{a^2}{8},$$

so that the equation (2.36) produces

$$h(s) = -\frac{4s}{3\pi} \sqrt{a^2 - s^2}.$$

Thus $f(t)$ is obtained by using (2.29).

(ii) Next, we consider the pair

$$\int_0^\infty f(t)\, J_1(xt)\, dt \;=\; \frac{ax}{\alpha},\ 0 \le x < 1, \tag{2.39}$$

$$\int_0^\infty \sqrt{t^2 + \frac{\alpha^2}{\beta^2}}\, f(t)\, J_1(xt)\, dt \;=\; 0,\ x > 1, \tag{2.40}$$

where a, α and β are non-zero real constants. This pair of integral equations arises in the study of the dynamic Reissner–Sagoci problem (cf. Shail 1970, Gladwell and Low 1970).

After setting

$$f(t) = \sqrt{t^2 + \frac{\alpha^2}{\beta^2}}\, g(t), \tag{2.41}$$

the above dual integral equations can be rewritten as

$$\int_0^\infty \sqrt{t^2 + \frac{\alpha^2}{\beta^2}}\, g(t)\, J_1(xt)\, dt \;=\; \frac{ax}{\alpha},\ 0 \le x < 1, \tag{2.42}$$

$$\int_0^\infty (t^2 + \frac{\alpha^2}{\beta^2})\, g(t)\, J_1(xt)\, dt \;=\; 0,\ x > 1. \tag{2.43}$$

The equation (2.43) is now equivalent to the ODE

$$\frac{d^2 u}{dx^2} + \frac{1}{x}\frac{du}{dx} - \left(\frac{1}{x^2} + \frac{\alpha^2}{\beta^2}\right) u(x) = 0,\ x > 1, \tag{2.44}$$

where

$$u(x) = \int_0^\infty g(t)\, J_1(xt)\, dt. \tag{2.45}$$

The solution of the ODE (2.44), keeping in mind that $u(x)$ is bounded at $x = \infty$, is

$$u(x) = A_1\, K_1\left(\frac{\alpha x}{\beta}\right),\ x > 1, \tag{2.46}$$

where A_1 is an arbitrary constant to be determined. From equations (2.45) and (2.46), we have

$$\int_0^\infty g(t)\, J_1(xt)\, dt = A_1\, K_1\left(\frac{\alpha x}{\beta}\right),\ x > 1. \tag{2.47}$$

Then, the dual integral equations (2.42) and (2.43) can be recast into the form (2.42) and (2.47). To obtain the solution of this pair, we now set

$$\int_0^\infty \sqrt{t^2 + \frac{\alpha^2}{\beta^2}}\, g(t)\, J_1(xt)\, dt = h(x),\ x > 1. \tag{2.48}$$

By using the Hankel inversion theorem, the equations (2.42) and (2.48) give

$$\sqrt{t^2 + \frac{\alpha^2}{\beta^2}} \; g(t) = \frac{a}{\alpha} \; J_2(t) + t \int_1^\infty s \; h(s) \; J_1(st) \; ds. \tag{2.49}$$

Substituting the above expression for $g(t)$ into the equation (2.47), an integral equation for the unknown function $h(s)$ is obtained as given by

$$\int_1^s s \; h(s) \; R(x, s) \; ds = A_1 \; K_1 \left(\frac{\alpha x}{\beta} \right) - \frac{a}{\alpha} \int_0^\infty \frac{J_2(t) \; J_1(xt)}{\sqrt{t^2 + \frac{\alpha^2}{\beta^2}}} \; dt, \; x > 1, \tag{2.50}$$

where

$$R(x, s) = \int_0^\infty \frac{t \; J_1(xt) \; J_1(st)}{\sqrt{t^2 + \frac{\alpha^2}{\beta^2}}} \; dt. \tag{2.51}$$

The solution of the integral equation (2.50) was obtained by Shail (1970) for large α. The asymptotic result for $R(x, s)$ as $\alpha \to \infty$ can be expressed as

$$R(x, s) \sim \frac{1}{\pi \sqrt{xs}} \; e^{\gamma(x+s)} \int_{\max(x,s)}^\infty \frac{e^{-2\gamma v} \; dv}{\sqrt{(v-x)(v-s)}} + O(\gamma^{-2}), \tag{2.52}$$

where $\gamma = \frac{\alpha}{\beta}$.

Substituting the expression for $R(x, s)$ from the relation (2.52) into equation (2.50), and interchanging the order of integration, we obtain an Abel integral equation of the second kind

$$\frac{1}{\pi} \int_x^\infty \frac{e^{-2\gamma v}}{\sqrt{v-x}} \left\{ \int_1^\infty \frac{\sqrt{s} \; e^{\gamma s} \; h(s)}{\sqrt{v-s}} \; ds \right\} dv = \sqrt{x} \; e^{-\gamma x} \; F(x), \; x > 1, \tag{2.53}$$

where

$$F(x) = A_1 \; K_1 \left(\frac{\alpha x}{\beta} \right) - \frac{a}{\alpha} \int_0^\infty \frac{J_2(t) \; J_1(xt)}{\sqrt{t^2 + \frac{\alpha^2}{\beta^2}}} \; dt.$$

We neglect the terms of $O(\gamma^{-2})$ in equation (2.53) because they are small. After repeated application of Abel inversion theorems, the solution of the integral equation (2.53) is obtained as given by

$$\sqrt{s} \; e^{\gamma s} \; h(s) = \frac{1}{\pi} \frac{d}{ds} \int_1^s \frac{e^{2\gamma v}}{\sqrt{s-v}} \left\{ \frac{d}{dv} \int_v^\infty \frac{\sqrt{x} \; e^{-\gamma x} \; F(x)}{\sqrt{x-v}} \; dx \right\} dv. \tag{2.54}$$

The constant A_1 in the relation (2.54) is determined from the condition that $h(1) = 0$. The relations (2.41), (2.49) and (2.54) give the complete solution of the dual integral equations (2.39) and (2.40) for large α.

25

(iii) Finally, we consider the dual integral equations

$$\int_0^\infty f(t)\, J_1(xt)\, dt \;=\; \omega x, \; 0 \le x \le a, \tag{2.55}$$

$$\int_0^\infty (\lambda t + \mu t^2)\, f(t)\, J_1(xt)\, dt \;=\; 0, \; x > a. \tag{2.56}$$

The above pair arises in the study of a viscous flow problem (cf. Goodrich 1969) induced by a rotating circular disc, kept on the surface of a bulk fluid of viscosity λ, which is otherwise contaminated by an adsorbed fluid film of different viscosity μ. ω is the constant angular velocity of the disc of radius a, rotating around its axis. For the cases (I) $\lambda = 0$, (II) $\mu = 0$ these equations possess explicit solutions. However, for the case when $\lambda \ne 0, \mu \ne 0$, Goodrich (1969) used a special method to solve the dual integral equations approximately. But as pointed out by Shail (1978), Goodrich's solutions are not the correct ones, since they involve certain divergent integrals. Chakrabarti (1989) solved this pair for the three cases in a straightforward manner as described below.

Case (I): $\lambda = 0$

In this case the dual integral equations (2.55), (2.56) reduce to the form

$$\int_0^\infty f(t)\, J_1(xt)\, dt \;=\; \omega x, \; 0 \le x \le a, \tag{2.57}$$

$$\int_0^\infty t^2\, f(t)\, J_1(xt)\, dt \;=\; 0, \; x > a. \tag{2.58}$$

The equation (2.58) is equivalent to the ODE

$$x^2 \frac{d^2 u}{dx^2} + x \frac{du}{dx} - u(x) = 0, \; x > a, \tag{2.59}$$

where

$$u(x) = \int_0^\infty f(t)\, J_1(xt)\, dt. \tag{2.60}$$

The solution of the ODE (2.59) is

$$u(x) = \frac{A_0}{x}, \; x > a, \tag{2.61}$$

using the condition that $u(x)$ is bounded at infinity, where A_0 is an arbitrary constant. Then

$$\int_0^\infty f(t)\, J_1(xt)\, dt = \frac{A_0}{x}, \; x > a. \tag{2.62}$$

26

To determine the constant A_0, we use the fact that $u(x)$ is continuous at $x = a$. This produces $A_0 = \omega a^2$.

Using the Hankel inversion theorem, equations (2.57) and (2.62) produce

$$f(t) = \omega a^2 \left[J_0(at) + J_2(at) \right]. \tag{2.63}$$

Case (II): $\mu = 0$

In this case, the pair is

$$\int_0^\infty f(t) \, J_1(xt) \, dt = \omega x, \ 0 \le x \le a, \tag{2.64}$$

$$\int_0^\infty t \, f(t) \, J_1(xt) \, dt = 0, \ x > a. \tag{2.65}$$

To find the solution of this pair, we assume that

$$\int_0^\infty t \, f(t) \, J_1(xt) \, dt = g(x), \ 0 \le x \le a, \tag{2.66}$$

where $g(x)$ is an unknown function to be determined. Using the Hankel inversion theorem, from equations (2.65) and (2.66), we obtain

$$f(t) = \int_0^a s \, g(s) \, J_1(st) \, ds. \tag{2.67}$$

Substituting the expression for $f(t)$ from (2.67) into (2.64) and then using the result (A.1.15), we get

$$\frac{2}{\pi x} \int_0^x \frac{v^2}{\sqrt{x^2 - v^2}} \left\{ \int_v^a \frac{g(s) \, ds}{\sqrt{s^2 - v^2}} \right\} dv = \omega x, \ 0 \le x \le a. \tag{2.68}$$

After repeated application of Abel's inversion theorems, from (2.68), we obtain

$$\pi \, g(x) = 4\omega x \, (a^2 - x^2)^{-1/2}, \ 0 \le x \le a. \tag{2.69}$$

Hence, the solution of the pair (2.64) and (2.65) is obtained from the relations (2.67) and (2.69).

Case (III): $\lambda \neq 0$, $\mu \neq 0$.

In this case, the most general pair of integral equations (2.55) and (2.56) is reduced to the solution of a Fredholm integral equation of the second kind. The method is given below.

We rewrite the equation (2.56) as

$$\int_0^\infty t \, (1 + k \, at) \, f(t) \, J_1(xt) \, dt = 0, \ x > a, \tag{2.70}$$

where $k = \dfrac{\mu}{\lambda a}$. Now, we put

$$\int_0^\infty t \, (1 + k \, at) \, f(t) \, J_1(xt) \, dt = g(x), \ 0 \leq x \leq a. \tag{2.71}$$

Then, by the Hankel inversion theorem, from (2.70) and (2.71), we find that

$$f(t) = \frac{1}{(1 + k \, at)} \int_0^a s \, g(s) \, J_1(st) \, ds. \tag{2.72}$$

Using the above relation in equation (2.55), we get the following integral equation for the unknown function $g(x)$:

$$\int_0^a s \, g(s) \left\{ \int_0^\infty \frac{J_1(xt) \, J_1(t)}{(1 + k \, at)} \, dt \right\} ds = \omega x, \ 0 \leq x \leq a. \tag{2.73}$$

Utilizing the relation

$$\frac{1}{1 + k \, at} = \frac{1}{2} \left[1 + \frac{1 - k \, at}{1 + k \, at} \right],$$

and the result (A.1.15) with repeated application of Abel's inversion theorems, the integral equation of the first kind (2.73) is reduced to the Fredholm integral equation of the second kind as given by

$$g^*(v) + \int_0^a K(u, v) \, g^*(u) \, du = 4\omega v, \ 0 \leq v \leq a, \tag{2.74}$$

where

$$g^*(v) \quad = \quad v \int_v^a \frac{g(s) \, ds}{\sqrt{s^2 - v^2}},$$

$$K(u, v) \quad = \quad \frac{2}{\pi} \int_0^\infty \frac{1 - k \, at}{1 + k \, at} \sin ut \, \sin vt \, dt.$$

28

A slightly more general dual integral equation than the pair (2.55) and (2.56), is

$$\int_0^\infty \left\{ 1 + \frac{1 - k\ at}{1 + k\ at}\ e^{-2ht} \right\} f(t)\ J_1(xt)\ dt\ =\ \omega x,\ 0 \le x \le a, \qquad (2.75)$$

$$\int_0^\infty t\ f(t)\ J_1(xt)\ dt\ =\ 0,\ x > a, \qquad (2.76)$$

where $h > 0$, which can also be reduced to a Fredholm integral equation of the second kind by the method described above. The integral equation pair (2.75) and (2.76) arises in the rotating disc problem of Shail (1979), when the disc is kept at a distance a below the contaminated surface considered by Goodrich (1969), so that the case $h = 0$ corresponds to the dual integral equations (2.55) and (2.56).

2.1.4 Now we present the method and solution of the following two pairs of integral equations involving $J_\nu(x)$:

$$\int_0^\infty t^2\ f(t)J_\nu(xt)\ dt\ =\ -g(x),\ 0 \le x \le 1, \qquad (2.77)$$

$$\int_0^\infty t\ f(t)\ J_\nu(xt)\ dt\ =\ 0,\ 1 \le x < \infty, \qquad (2.78)$$

and

$$\int_0^\infty f(t)\ J_\nu(xt)\ dt\ =\ g(x),\ 0 \le x \le 1, \qquad (2.79)$$

$$\int_0^\infty (\lambda t + \mu t^2) f(t)\ J_\nu(xt)\ dt\ =\ 0,\ 1 \le x < \infty. \qquad (2.80)$$

For $\nu = 1$, these pairs have already been discussed in §2.1.3. The above general pairs are studied by Srivastava and Srivastava (1991). Solutions of these dual integral equations are obtained by the same method as described in §2.1.3.

To find the solution of the dual integral equations (2.77) and (2.78), we assume that

$$\int_0^\infty t\ f(t)\ J_\nu(xt)\ dt = \phi(x),\ 0 \le x \le 1. \qquad (2.81)$$

Then, by the Hankel inversion theorem, equations (2.78) and (2.81) give

$$f(t) = \int_0^1 s\ \phi(s)\ J_\nu(st)\ ds. \qquad (2.82)$$

29

Substituting the expression for $f(t)$ from (2.82) into equation (2.77), we get

$$\int_0^\infty t^2 \, J_\nu(xt) \left\{ \int_0^1 s \, \phi(s) J_\nu(st) \, ds \right\} dt = -g(x), \ 0 \le x \le 1. \qquad (2.83)$$

The above equation is equivalent to the ODE

$$\frac{d^2u}{dx^2} + \frac{1}{x}\frac{du}{dx} - \frac{\nu^2}{x^2} \, u(x) = g(x), \ 0 \le x \le 1, \qquad (2.84)$$

where

$$u(x) = \int_0^\infty J_\nu(xt) \left\{ \int_0^1 s \, \phi(s) \, J_\nu(st) \, ds \right\} dt. \qquad (2.85)$$

The solution of the ODE (2.84), by the method of variation of parameters, is

$$u(x) = A \, x^\nu + \frac{B}{x^\nu} + \frac{x^\nu}{2\nu} \int_0^x s^{1-\nu} g(s) \, ds - \frac{1}{2\nu x^\nu} \int_0^x s^{1+\nu} g(s) \, ds, \qquad (2.86)$$

for $\nu > 0$ and

$$u(x) = A + B \log x + \int_0^x \frac{1}{t} \left\{ \int_0^t s \, g(s) \, ds \right\} dt \qquad (2.87)$$

for $\nu = 0$, where A and B are arbitrary constants to be determined from physical considerations of the problem. The constant B must be zero since $u(x)$ is required to be finite at $x = 0$. The other constant A will be determined later.

Using the result (A.1.15), then interchanging the order of integration, from (2.85) we find that

$$\frac{2}{\pi} \int_0^x \frac{v^{2\nu}}{\sqrt{x^2 - v^2}} \left\{ \int_v^1 \frac{s^{1-\nu} \, \phi(s)}{\sqrt{s^2 - v^2}} \, ds \right\} dv = x^\nu \, u(x). \qquad (2.88)$$

By Abel's inversion theorem, the above equation produces another Abel integral equation

$$\int_v^1 \frac{s^{1-\nu} \, \phi(s)}{\sqrt{s^2 - v^2}} \, ds = v^{1-2\nu} \int_0^v \frac{\frac{d}{dx}\{x^\nu u(x)\}}{\sqrt{v^2 - x^2}} \, dx = \psi(v), \quad \text{say.} \qquad (2.89)$$

By another use of Abel's inversion theorem, we get

$$s^{1-\nu}\phi(s) = \frac{2s}{\pi} \left\{ \frac{\psi(1)}{\sqrt{1 - s^2}} - \int_s^1 \frac{\psi'(t)}{\sqrt{t^2 - s^2}} \, dt \right\}. \qquad (2.90)$$

Equations (2.82) and (2.90) give the complete solution of the dual integral equations (2.77) and (2.78) provided the constant A is determined. To find the constant A, we use the relation (2.90) and the observation that $\psi(1) = 0$ arising from the continuity

30

requirement of $\phi(x)$ at $x = 1$. Therefore, for $\nu > 0$, the constant A is determined by the relation

$$\int_0^1 \frac{\frac{d}{dx}\{x^\nu u(x)\}}{\sqrt{1-x^2}} \, dx = 0. \tag{2.91}$$

For $\nu = 0$, equation (2.88) can be used to determine the constant A.

As in §2.1.3, the solution of the dual integral equations (2.79) and (2.80) is obtained in three cases.

Case (I): $\lambda = 0$

Following a similar analysis to that presented in §2.1.3, the solution of the pair

$$\int_0^\infty f(t) \, J_\nu(xt) \, dt \;=\; g(x), \; 0 \le x \le 1, \tag{2.92}$$

$$\int_0^\infty t^2 \, f(t) \, J_\nu(xt) \, dt \;=\; 0, \; 1 \le x < \infty, \tag{2.93}$$

is given by

$$f(t) = g(1) \, J_{\nu-1}(t) + t \int_0^1 x \, g(x) \, J_\nu(xt) \, dx. \tag{2.94}$$

Case (II): $\mu = 0$

In a similar manner, to obtain the solution of the dual integral equations

$$\int_0^\infty f(t) \, J_\nu(xt) \, dt \;=\; g(x), \; 0 \le x \le 1, \tag{2.95}$$

$$\int_0^\infty t \, f(t) \, J_\nu(xt) \, dt \;=\; 0, \; 1 \le x < \infty, \tag{2.96}$$

we set

$$\int_0^\infty t \, f(t) \, J_\nu(xt) \, dt = \phi(x), \; 0 \le x \le 1. \tag{2.97}$$

By the Hankel inversion theorem, from (2.96) and (2.97), we get

$$f(t) = \int_0^1 x \, \phi(x) \, J_\nu(xt) \, dx. \tag{2.98}$$

Substituting this value of $f(t)$ from (2.98) into equation (2.95) and proceeding as in the previous analysis for the pair (2.77) and (2.78), we get the integral equation (2.88) with $g(x)$ in place of $u(x)$ whose solution is already obtained.

Case (III): $\lambda \neq 0$ and $\mu \neq 0$

Following a similar analysis as is §2.1.3, the pair of integral equations (2.79) and (2.80) can be reduced to the solution of a Fredholm integral equation of the second kind

$$\frac{f(t)}{t} - \lambda \int_0^\infty K(x,t) \, \frac{f(x)}{x} \, dx = G(t), \tag{2.99}$$

where

$$G(t) = \frac{\mu}{\lambda + \mu t} \left\{ g(1) \, J_{\nu-1}(t) + t \int_0^1 s \, g(s) \, J_\nu(st) \, ds \right\},$$

and

$$K(x,t) = \frac{1}{(\lambda + \mu t)} \left\{ J_{\nu-1}(t) J_\nu(x) + \frac{xt J_{\nu+1}(x) J_\nu(t) - t^2 J_{\nu+1}(t) J_\nu(x)}{x^2 - t^2} \right\}.$$

Finally, we consider the generalization of the pair (2.42) and (2.43) as

$$\int_0^\infty \sqrt{t^2 + r^2} \, f(t) \, J_\nu(xt) \, dt = g(x), \ 0 \le x \le 1, \tag{2.100}$$

$$\int_0^\infty (t^2 + r^2) \, f(t) \, J_\nu(xt) \, dt = 0, \ 1 \le x < \infty, \tag{2.101}$$

where r is a real non-zero constant.

Equation (2.101) is equivalent to the ODE

$$\frac{d^2 u}{dx^2} + \frac{1}{x} \frac{du}{dx} - \left(\frac{\nu^2}{x^2} + r^2 \right) u(x) = 0, \ 1 \le x < \infty, \tag{2.102}$$

where

$$u(x) = \int_0^\infty f(t) \, J_\nu(xt) \, dt.$$

The solution of the ODE (2.102) is

$$u(x) = A \, K_\nu(xr) + B \, I_\nu(xr), \ 1 \le x < \infty.$$

Since $u(x)$ must be finite at infinity, B must be zero and then we get

$$\int_0^\infty f(t) \, J_\nu(xt) \, dt = A \, K_\nu(xr), \ 1 \le x < \infty. \tag{2.103}$$

Now we assume that

$$\int_0^\infty \sqrt{t^2 + r^2} \, f(t) \, J_\nu(xt) \, dt = \phi(x), \ 1 \le x < \infty. \tag{2.104}$$

32

Then, from equations (2.100) and (2.104), by the Hankel inversion theorem, we have

$$\sqrt{t^2 + r^2}\ f(t) = t \left\{ \int_0^1 s\ g(s)\ J_\nu(st)\ ds + \int_1^\infty s\ \phi(s)\ J_\nu(st)\ ds \right\}. \tag{2.105}$$

Substituting the expression for $f(t)$ from (2.105) into equation (2.103), we obtain

$$\int_1^\infty s\ L(s, x)\ \phi(s)\ ds = G(x), \tag{2.106}$$

where

$$L(s, x) = \int_0^\infty t\ (t^2 + r^2)^{-1/2}\ J_\nu(st)\ J_\nu(xt)\ dt$$

and

$$G(x) = A\ K_\nu(xr) - \int_0^1 s\ g(s)\ L(s, x)\ ds.$$

The constant A is determined from the continuity requirement at $x = 1$ which gives $\phi(1) = g(1)$.

As in §2.1.3, the solution of the integral equation (2.106) is derived for large r for which we use the asymptotic result for $L(s, x)$ as given by

$$L(s, x) \sim \frac{1}{\pi (sx)^{1/2}}\ e^{r(s+r)} \int_{\max(s,x)}^\infty \frac{e^{-2rv} dv}{\sqrt{(v-s)(v-x)}} + O\left(\frac{1}{r^2}\right).$$

Using the above result in equation (2.106) and then after some simplification, we obtain

$$s^{1-\frac{\nu}{2}}\ e^{rs}\ \phi(s) = \frac{1}{\pi} \frac{d}{ds} \int_1^s \frac{e^{2rv}}{\sqrt{s-v}} \left\{ \frac{d}{dv} \int_v^\infty \frac{x^{\nu/2}\ e^{-rx}\ G(x)}{\sqrt{x-v}}\ dx \right\}\ dv. \tag{2.107}$$

The solution of the dual integral equations is then obtained by using the relation (2.105).

2.1.5 Here we consider the dual integral equations (cf. Rose and De Hoog 1983)

$$\int_0^\infty t^{2\alpha}\ \{1 + \omega(t)\}\ f(t)\ (xt)^{1/2}\ J_\nu(xt)\ dt\ =\ g(x),\ 0 \le x < a, \tag{2.108}$$

$$\int_0^\infty f(t)\ (xt)^{1/2}\ J_\nu(xt)\ dt\ =\ 0,\ a \le x < \infty, \tag{2.109}$$

where $\omega(t)$ is a known weight function. The above pair of integral equations arises in the solution of the following mixed boundary value problem.

33

The problem is to find the solution of the two-dimensional Laplace equation

$$\frac{\partial^2 u}{\partial x^2} + \frac{\partial^2 u}{\partial y^2} = 0, \ 0 \le x < \infty, \ 0 \le y \le b, \tag{2.110}$$

satisfying the boundary conditions

$$\frac{\partial u}{\partial x}(0, y) = 0 \tag{2.111}$$

$$u(x, b) = 0, \ 0 \le x < \infty, \tag{2.112}$$

$$u(x, 0) = 0, \ x \ge a, \tag{2.113}$$

$$\frac{\partial u}{\partial x}(x, 0) = -g(x), \ 0 \le x \le a, \tag{2.114}$$

$$u(x, y) = 0, \ x \to \infty. \tag{2.115}$$

The solution of this problem can be continued as an even function of x and an odd function of y into the strip $|x| < \infty$, $|y| \le b$, with a cut along $|x| \le a$, $y = 0$ across which u has a jump discontinuity. In this extended domain, $u(x, y)$ could be interpreted as the elastic displacement for an anti-plane crack problem and the derivatives $\frac{\partial u}{\partial x}, \frac{\partial u}{\partial y}$ as the only non-vanishing components of the stress tensor (provided we choose the shear modulus as the unit of stress). The cut along $|x| \le a$, $y = 0$ would correspond to the crack and the boundary condition (2.114) would specify the tractions applied to the crack faces. We consider the integral representation for the solution as

$$u(x, y) = \sqrt{\frac{2}{\pi}} \int_0^\infty U(t) \sinh t(b - y) \ \cos xt \ dt, \tag{2.116}$$

for $0 \le x < \infty$, $0 \le y < b$, where $U(t)$ is an unknown function to be determined. This representation satisfies (2.110) and the boundary conditions (2.111) and (2.112). To satisfy the other boundary conditions, the unknown function $U(t)$ must satisfy the dual integral equations

$$\sqrt{\frac{2}{\pi}} \int_0^\infty t \, U(t) \, \cosh bt \ \cos xt \ dt = g(x), \ 0 \le x < a, \tag{2.117}$$

$$\sqrt{\frac{2}{\pi}} \int_0^\infty U(t) \, \sinh bt \ \cos xt \ dt = 0, \ a \le x < \infty. \tag{2.118}$$

34

Putting $f(t) = \sinh bt \, U(t)$, $\nu = -\frac{1}{2}$, $\alpha = \frac{1}{2}$, $1 + \omega(t) = \cosh bt$, we observe that the above pair is a particular case of the pair (2.108) and (2.109). Following Noble (1963), the pair (2.108) and (2.109) can be reduced to a Fredholm integral equation of the second kind

$$\psi(x) + \int_0^a t \, K(x,t) \, \psi(t) \, dt = G_1(x), \qquad (2.119)$$

with

$$K(x,t) = \int_0^\infty s \, \omega(s) \, J_{\nu+\alpha}(xs) \, J_{\nu+\alpha}(ts) \, ds, \qquad (2.120)$$

$$G_1(x) = \frac{2^{1-\alpha} x^{-\nu-\alpha}}{\Gamma(\alpha)} \int_0^x s^{\frac{1}{2}+\nu} \, (x^2 - s^2)^{\alpha-1} \, g(s) \, ds, \qquad (2.121)$$

where

$$f(t) = t^{\frac{1}{2}-\alpha} \int_0^\infty s \, J_{\nu+\alpha}(st) \, \psi(s) \, ds, \qquad (2.122)$$

$Re \; \nu \geq \frac{1}{2}$, $0 < Re \; \alpha < 1$, $\omega(s) = o(1)$ as $s \to \infty$, $\omega(s) = O(s^\gamma)$ as $s \to 0$, $\gamma > -2(1 + \nu + \alpha)$.

Thus the dual integral equations (2.117) and (2.118) can be reduced to the second-kind Fredholm integral equation (2.119). However, as the mixed boundary value problem can be solved exactly by complex-variable techniques, Rose and De Hoog (1983) stated that this 'constitutes a useful source of exact solutions of dual integral equations'. This method is now described below.

The basic idea of the complex-variable approach is to map conformally the original domain in the complex z-plane ($z = x + iy$) onto a domain in the ζ-plane ($\zeta = \xi + i\eta$) which is more convenient for the analysis. The most efficient choice for the mapping function in any particular case depends on the particular set of boundary conditions involved. A convenient choice for problem (2.110)−(2.115) is the transformation

$$\zeta = \tanh\left(\frac{\pi z}{2b}\right), \qquad (2.123)$$

which maps the strip $|x| < \infty$, $|y| \leq b$ onto the whole ζ-plane with two branch cuts along $|\xi| \geq 1$, $\eta = 0$. The half-lines $0 \leq x < \infty$, $y = b$ and $0 \leq x < \infty$, $y = -b$

are mapped respectively onto the upper bank and the lower bank of the branch cut $\xi \geq 1$, $\eta = 0$.

Defining a function $w(\xi, \eta)$ in the half-plane $\eta \geq 0$ by

$$w(\xi, \eta) = u(x, y), \tag{2.124}$$

where $u(x, y)$ is the solution of the problem continued into $x < 0$ as an even function of x, and (ξ, η) is the image of the point (x, y) under the transformation (2.123), we see that $w(\xi, \eta)$ satisfies

$$\frac{\partial^2 w}{\partial \xi^2} + \frac{\partial^2 w}{\partial \eta^2} = 0, \ \eta > 0, \tag{2.125}$$

with the boundary conditions

$$w = 0; \ \beta \leq |\xi| < \infty, \ \eta \to 0+, \ \beta = \tanh\left(\frac{\pi a}{2b}\right), \tag{2.126}$$

$$\frac{\partial w}{\partial \eta} = -h(\xi); \ 0 \leq |\xi| < \beta, \ \eta \to 0+, \tag{2.127}$$

where

$$h(\xi) = \frac{2b}{\pi(1 - \xi^2)} \ g\left(\frac{2b}{\pi} \tanh^{-1} \xi\right), \tag{2.128}$$

and

$$w \to 0 \ \text{as} \ \xi^2 + \eta^2 \to \infty. \tag{2.129}$$

Equations (2.125)−(2.129) represent a two-part mixed boundary value problem whose solution is obtained as follows.

We set

$$W(\zeta) = w(\xi, \eta) + i \chi(\xi, \eta) \tag{2.130}$$

$$= \frac{1}{\pi i} \int_{-\beta}^{\beta} \frac{\lambda(t)}{\sqrt{\zeta^2 - t^2}} \, dt, \tag{2.131}$$

where w, χ are real, conjugate harmonic functions and $W(\zeta)$ is the corresponding complex potential involving an unknown real function $\lambda(t)$ which is an even function of t for $|t| \leq \beta$. The principal branch of $\sqrt{\zeta^2 - t^2}$ is chosen which is real for $|\xi| \geq t$, $\eta = 0$. The boundary conditions (2.126) and (2.129) are automatically satisfied.

From (2.130), we get

$$W'(\zeta) = \frac{\partial w}{\partial \xi} - i\,\frac{\partial w}{\partial \eta}. \tag{2.132}$$

From (2.131), when $\zeta = \xi + i.0+$,

$$W'(\zeta) = \frac{\partial W}{\partial \xi} = \frac{d}{d\xi}\left\{ \frac{1}{\pi i}\int_0^\xi \frac{\lambda(t)}{\sqrt{\xi^2 - t^2}}\,dt - \frac{1}{\pi}\int_\xi^\beta \frac{\lambda(t)}{\sqrt{t^2 - \xi^2}}\,dt \right\}. \tag{2.133}$$

Hence, the boundary condition (2.127) produce an Abel integral equation of the first kind for the unknown function $\lambda(t)$ as

$$\frac{1}{\pi}\frac{d}{d\xi}\int_0^\xi \frac{\lambda(t)\,dt}{\sqrt{\xi^2 - t^2}} = -h(\xi),$$

whose solution is given by

$$\lambda(t) = -2t\int_0^t \frac{h(s)}{\sqrt{t^2 - s^2}}\,ds. \tag{2.134}$$

The complex potential $W(\zeta)$ can now be determined from the relations (2.128) and (2.134). Therefore, the solution of the problem is obtained from (2.124) and (2.130). After obtaining $u(x, y)$ and then applying the Fourier inversion theorem to (2.116) together with the use of the fact that $f(t) = \sinh bt\ U(t)$, the exact solution of the dual integral equations (2.108) and (2.109) is found.

In particular, from (2.124), (2.130) and (2.131), we find that

$$u(x, 0+) = Re\left[W(\xi + i.0+)\right],\ 0 \le x < a,\ y \to 0+$$

$$= \frac{2}{\pi}\int_\xi^\beta \frac{t}{\sqrt{t^2 - \xi^2}}\left\{ \int_0^t \frac{h(s)\,ds}{\sqrt{t^2 - s^2}} \right\}\,dt,$$

$$\frac{\partial u}{\partial y}(x, 0+) = -\frac{2}{\pi}\frac{d}{dx}\int_0^\beta \frac{t}{\sqrt{\xi^2 - t^2}}\left\{ \int_0^t \frac{h(s)\,ds}{\sqrt{t^2 - s^2}} \right\}\,dt,\ x > a.$$

In applications dealing with crack problems, it is not usually necessary to know the displacement $u(x, y)$ throughout the elastic medium. It is required to find the value of the stress intensity factor $S(a)$ defined by

$$S(a) = \lim_{x \to a-}\left[\{2(x - a)\}^{-\frac{1}{2}}\,u(x, 0+) \right]$$

$$= \lim_{x \to a+}\left[\{2(a - x)\}^{\frac{1}{2}}\,\frac{\partial u}{\partial y}(x, 0+) \right].$$

37

Rose and De Hoog (1983) also studied two other mixed boundary value problems for crack problems which can also be solved explicitly after reducing it to the solution of appropriate dual integral equations by the above analysis.

2.1.6 Anderssen et al. (1982) obtained an explicit solution to the dual integral equations (2.108) and (2.109) when $\omega(t)$ takes the form

$$\omega(t) = \frac{k}{t^2 + c^2}; \ k, c \ \text{are constants.} \tag{2.135}$$

Substituting the above expression for $\omega(t)$ into the relation (2.120) with $\nu + \alpha = 0$, then using the result (A.1.17), the kernel $K(x,t)$ reduces to the form

$$K(x,t) = k \ G(x,t;c),$$

where

$$G(x,t;c) = \begin{cases} I_0(ct) \ K_0(cx), & 0 < t < x, \\ I_0(cx) \ K_0(ct), & t > x. \end{cases} \tag{2.136}$$

Now, we introduce the transformation

$$\Phi(x;c) = \int_0^a t \ G(x,t;c) \ \phi(t) \ dt, \tag{2.137}$$

where $\phi \in C[0,a]$, and define the operator L_p by

$$L_p u = \frac{1}{x} \frac{d}{dx}\left(x \frac{du}{dx} \right) - \left(\frac{p}{x}\right)^2 u, \ p \geq 0. \tag{2.138}$$

Lemma : If $\phi \in C[0,a]$, then $\Phi \in C^2[0,a]$,

$$(L_0 - c^2 I)\Phi(x;c) \ = \ -\phi(x), \tag{2.139}$$

$$\Phi'(0;c) \ = \ 0, \tag{2.140}$$

$$c \ K_0'(ca) \ \Phi(a;c) - K_0(ca) \ \Phi'(a;c) \ = \ 0, \tag{2.141}$$

where I denotes the identity operator.

From the above lemma, we conclude that $t \ G(x,t;c)$ is a Green's function of the operator $-(L_0 - c^2 I)$.

38

Theorem If $\omega(t) = \dfrac{k}{t^2 + c^2}$, $\nu + \alpha = 0$ and $G_1(x) \in C[0, a]$, the general solution of the Fredholm integral equation of the seond kind (2.119) is given by

$$\psi(x) = G_1(x) - k\, Q(x; c_1) + A\, Q(a; c_1)\, I_0(c_1 x)/I_0(c_1, a), \qquad (2.142)$$

where

$$Q(x; c_1) = \int_0^a t\, G(x, t; c_1)\, G_1(t)\, dt, \; c_1 = \sqrt{c^2 + k}, \qquad (2.143)$$

and

$$A = \frac{k\,\{c - c_1 K_0(ca) K_1(c_1 a)/K_1(ca) K_0(c_1 a)\}}{\{c + c_1 K_0(ca) I_1(c_1 a)/K_1(ca) I_0(c_1 a)\}}. \qquad (2.144)$$

Proof It follows from (2.142) that $(\psi(x) - G_1(x)) \in C^2[0, a]$. Defining

$$\Psi(x; c) = \int_0^a t\, G(x, t; c)\, \psi(t)\, dt,$$

we see from (2.139) that $t\, G(x, t; c_1)$ is a Green's function of the operator $-(L_0 - (c^2 + k)I)$. Using the results and noting that $I_0(c_1 x)$ is in the null space of the operator $(L_0 - (c^2 + k)I)$, we find that

$$(L_0 - (c^2 + k)I)(\psi - G_1)(x) = k\, G_1(x).$$

This can be rewritten as

$$(L_0 - c^2 I)\,\{\psi(x) + k\, \Psi(x; c) - G_1(x)\} = 0. \qquad (2.145)$$

We have to show that $\psi(x)$ satisfies

$$\psi(x) - G_1(x) + k\, \Psi(x; c) = 0, \qquad (2.146)$$

which is the integral equation defined by (2.119), (2.120) and (2.135). From the condition (2.141), the integral equation (2.146) gives

$$c\, K_0'(ca)\, (\psi(a) - G_1(a)) - K_0(ca)\, (\psi'(a) - G_1'(a)) = 0.$$

Also from (2.141), we have

$$c_1\, K_0'(c_1 a)\, Q(a; c_1) - K_0(c_1 a)\, Q'(a; c_1) = 0.$$

Thus, the above two relations and (2.142) gives the expression (2.144) for the constant A. We also see that equation (2.142) satisfies the condition (2.140). Therefore, the explicit solution to the dual integral equations (2.108) and (2.109) is obtained from the relations (2.122) and (2.142).

Anderssen et al. (1982) also used the same technique to derive explicit solution for more general forms of $\omega(t)$ such as

$$\omega(t) = \sum_{i=1}^{n} \frac{k_i}{(t^2 + c_i^2)}, \ \sum_{i=1}^{n} \frac{k_i}{(t^2 + c_i^2)^i},$$

and linear combinations of such sums.

2.1.7 Dual integral equations involving Bessel functions of the first kind of the same order as kernels were considered by several researchers and which are presented systematically in the treatise by Sneddon (1966). Here, we study some dual integral equations with Bessel functions of the first kind of different orders as kernels. Nasim and Aggarwala (1984) considered the pair of dual integral equations

$$\int_0^\infty t^{-2\alpha} f(t) \, J_\nu(xt) \, dt \ = \ g(x), \ 0 < x < 1, \tag{2.147}$$

$$\int_0^\infty t^{-2\beta} f(t) \, J_\mu(xt) \, dt \ = \ h(x), \ 1 < x < \infty. \tag{2.148}$$

The solution of this pair is obtained explicitly by using Erdélyi–Kober fractional integral operators. To find the solution of the above pair, we consider two cases.

Case I: $h(x) = 0$

We consider the pair

$$\int_0^\infty t^{-2\alpha} f(t) \, J_\nu(xt) \, dt \ = \ g(x), \ 0 < x < 1, \tag{2.149}$$

$$\int_0^\infty t^{-2\beta} f(t) \, J_\mu(xt) \, dt \ = \ 0, \ 1 < x < \infty. \tag{2.150}$$

We write (2.150) as

$$\int_0^\infty t^{-2\beta - \frac{1}{2}} f(t) \, J_\mu(xt) \, (xt)^{1/2} \, dt = \phi(x) \, H(1 - x), \ \mu > -\frac{1}{2}, \tag{2.151}$$

40

where $\phi(x)$ is an unknown function and $H(x)$ is the Heaviside function. Then by the Hankel inversion theorem, equations (2.150) and (2.151) give

$$t^{-2\beta-\frac{1}{2}}f(t) = \int_0^1 \phi(x) \, J_\mu(xt) \, (xt)^{1/2} \, dx, \quad \mu > -\frac{1}{2}. \qquad (2.152)$$

Using the relation

$$J_\mu(xt) = \frac{2^{\eta-\mu+1}}{\Gamma(\mu-\eta)} x^{-\mu} \, t^{\mu-\eta} \int_0^x u^{1+\eta} \, (x^2 - u^2)^{\mu-\eta-1} \, J_\eta(ut) \, du$$

where $\eta = \frac{1}{2}\nu + \frac{1}{2}\mu - \alpha + \beta$ and $-1 < \eta < \mu$, in equation (2.152) and then changing the order of integration, we find that

$$t^{-2\beta-\frac{1}{2}} \, f(t) = \frac{2^{\eta-\mu+1}}{\Gamma(\mu-\eta)} t^{\mu-\eta+\frac{1}{2}} \int_0^1 u^{1+\eta} \, J_\eta(ut)$$

$$\times \left\{ \int_u^1 x^{\frac{1}{2}-\mu} \, \phi(x) \, (x^2 - u^2)^{\mu-\eta-1} dx \right\} du.$$

The above equation can be written as

$$f(t) = t^{2\beta+\mu-\eta+1} \int_0^1 \omega(u) \, J_\eta(ut) \, du, \quad \text{say.} \qquad (2.153)$$

Substituting this into equation (2.149), we get

$$g(x) = \int_0^\infty t^{\eta-\nu+1} J_\nu(xt) \left\{ \int_0^1 \omega(u) J_\eta(ut) \, du \right\} dt$$

$$= \int_0^1 \omega(u) \left\{ \int_0^\infty t^{\eta-\nu+1} \, J_\nu(xt) \, J_\eta(ut) \, dt \right\} du$$

$$= \frac{2^{\eta-\nu+1}}{\Gamma(\nu-\eta)} \, x^{-\nu} \int_0^x u^\eta \, (x^2 - u^2)^{\nu-\eta-1} \, \omega(u) \, du, \quad 0 < x < 1 \quad \text{(by A.1.18)},$$

where $\eta = \frac{1}{2}\mu + \frac{1}{2}\nu - \alpha + \beta$, $-1 < \eta < \mu$ and $-1 < \eta < \nu$.

The above expression can be expressed as

$$g(x) = 2^{\eta-\nu} \, I_{\eta-\frac{\nu}{2}, \, \nu-\eta} \left[x^{\nu-\eta-1}\omega(x) \right],$$

where I is the Erdélyi−Kober fractional integral operator (cf. Sneddon 1966, p. 48).

By the usual inversion, we have

$$\omega(x) = 2^{\nu-\eta}x^{\eta-\nu+1}I^{-1}_{\eta-\frac{\nu}{2},\ \nu-\eta}[g(x)]$$
$$= 2^{\nu-\eta}x^{\eta-\nu+1}I_{\frac{\nu}{2},\ \eta-\nu}[g(x)].$$

Therefore, from (2.153), we obtain

$$f(t) = 2^{\nu-\eta}\ t^{2\beta+\mu-\eta+1}\int_0^1 u^{\eta-\nu+1}J_\eta(ut)\ I_{\frac{\nu}{2},\ \eta-\nu}[g]\ du$$
$$\equiv f_1(t),\quad \text{say,} \tag{2.154}$$

which is the solution of the dual integral equations (2.149) and (2.150).

Case II: $g(x) = 0$

In this case, the pair is

$$\int_0^\infty t^{-2\alpha}\ f(t)\ J_\nu(xt)\ dt = 0,\ 0 < x < 1, \tag{2.155}$$

$$\int_0^\infty t^{-2\beta}\ f(t)\ J_\mu(xt)\ dt = h(x),\ 1 < x < \infty. \tag{2.156}$$

As in Case I, by Hankel inversion and for an appropriate unknown function $\phi(x)$, equation (2.155) produces

$$t^{-2\alpha-\frac{1}{2}}f(t) = \int_1^\infty \phi(x)\ J_\nu(xt)\ (xt)^{\frac{1}{2}}\ dx,\ \nu > -\frac{1}{2}.$$

Using the integral representation (cf. Erdélyi et al. 1954b, p. 25).

$$J_\nu(xt) = \frac{2^{\nu-\eta+1}}{\Gamma(\eta-\nu)}\ x^\nu\ t^{\eta-\nu}\int_x^\infty u^{1-\eta}(u^2-x^2)^{\eta-\nu-1}J_\eta(ut)\ du,$$

where $\eta = \frac{\nu}{2} + \frac{\mu}{2} - \alpha + \beta$ and $\nu < \eta < 2\nu + \frac{3}{2}$, in the above relation and changing the order of integration, we find

$$f(t) = \frac{2^{\nu-\eta+1}}{\Gamma(\eta-\nu)}t^{2\alpha+\eta-\nu+1}\int_1^\infty u^{1-\eta}\ J_\eta(ut)\left\{\int_1^u \phi(x)\ x^{\frac{1}{2}+\nu}(u^2-x^2)^{\eta-\nu-1}dx\right\}du$$
$$= t^{2\alpha+\eta-\nu+1}\int_1^\infty \omega(u)\ J_\eta(ut)\ du,\quad \text{say.} \tag{2.157}$$

42

Substituting the above expression into equation (2.156), we get

$$
\begin{aligned}
h(x) &= \int_0^\infty t^{\mu-\eta+1} J_\mu(xt) \left\{ \int_1^\infty \omega(u) \, J_\eta(ut) \, du \right\} \\[2mm]
&= \int_1^\infty \omega(u) \left\{ \int_0^\infty t^{\mu-\eta+1} J_\mu(xt) \, J_\eta(ut) \, dt \right\} du \\[2mm]
&= \frac{2^{\mu-\eta+1}}{\Gamma(\eta-\mu)} \, x^\mu \int_x^\infty u^\eta \, \omega(u) \, (u^2 - x^2)^{\eta-\mu-1} du, \quad 1 < x < \infty \quad \text{(by A.1.18)},
\end{aligned}
$$

where $-1 < \mu < \eta$ and $\nu < \eta < 2\nu + \frac{3}{2}$.

Using the Erdélyi–Kober operator K (cf. Sneddon 1966, p. 50), the above relation can be represented as

$$
h(x) = 2^{\mu-\eta} \, K_{\frac{\mu}{2}, \, \eta-\mu} \left[x^{\eta-\mu-1} \omega(x) \right]
$$

which by inversion gives

$$
\omega(x) = 2^{\eta-\mu} x^{\mu-\eta+1} \, K_{\eta-\frac{\mu}{2}, \, \mu-\eta}[h(x)].
$$

Thus, from (2.157) we get

$$
\begin{aligned}
f(t) &= 2^{\eta-\mu} \, t^{2\alpha+\eta-\nu+1} \int_1^\infty u^{\mu-\eta+1} \, J_\eta(ut) \, K_{\eta-\frac{\mu}{2}, \, \mu-\eta}[h] \, du \\[2mm]
&\equiv f_2(t), \quad \text{say},
\end{aligned} \tag{2.158}
$$

which is the solution of the pair of integral equations (2.155) and (2.156). Combining (2.154) and (2.158), we obtain the solution of the dual integral equations (2.147) and (2.148) as

$$
f(t) = f_1(t) + f_2(t). \tag{2.159}
$$

In particular, if $\eta - \nu > 0$ and $\mu - \eta > 0$ for which $\nu - \mu < 2(\beta - \alpha)$, then the relation (2.159) becomes

$$
\begin{aligned}
f(t) &= \frac{2^{\nu-\eta+1}}{\Gamma(\eta-\nu)} t^{2\beta+\mu-\eta+1} \int_0^1 x^{1-\eta} \, J_\eta(xt) \left\{ \int_0^x (x^2 - u^2)^{\eta-\nu-1} u^{1+\nu} g(u) \, du \right\} dx \\[2mm]
&\quad + \frac{2^{\eta-\mu+1}}{\Gamma(\mu-\eta)} t^{2\alpha+\eta-\nu+1} \int_1^\infty x^{1+\eta} \, J_\eta(xt) \left\{ \int_x^\infty \frac{(u^2-x^2)^{\mu-\eta-1}}{u^{\mu+1}} h(u) \, du \right\} dx.
\end{aligned}
$$

If $\mu = \nu$, then the solution of the dual integral equations

$$\int_0^\infty t^{-2\alpha} f(t) \, J_\nu(xt) \, dt \;=\; g(x), \; 0 < x < 1,$$
$$\int_0^\infty t^{-2\beta} \, f(t) \, J_\nu(xt) \, dt \;=\; h(x), \; 1 < x < \infty,$$

is given by

$$f(t) \;=\; 2^{\alpha-\beta} t^{1+\alpha+\beta} \int_0^1 x^{\beta-\alpha+1} \, J_{\nu+\beta-\alpha}(xt) \, I_{\frac{\nu}{2}, \, \beta-\alpha}[g] \; dx$$

$$+ 2^{\beta-\alpha} t^{1+\alpha+\beta} \int_1^\infty x^{\alpha-\beta+1} \, J_{\nu+\beta-\alpha}(xt) \, K_{\frac{\nu}{2}+\beta-\alpha, \; \alpha-\beta}[h] \; dx.$$

If $0 < \beta - \alpha < 1$, then the above solution becomes

$$f(t) \;=\; \frac{2^{1+\alpha-\beta} \, t^{1+\alpha+\beta}}{\Gamma(\beta-\alpha)} \int_0^1 x^{1+\alpha-\beta-\nu} J_{\nu+\beta-\alpha}(xt) \left\{ \int_0^x \frac{(x^2-u^2)^{\beta-\alpha-1}}{u^{-1-\nu}} g(u) \, du \right\} dx$$

$$- \frac{2^{\beta-\alpha}}{\Gamma(1+\alpha-\beta)} t^{1+\alpha+\beta} \int_1^\infty x^{\nu+\beta-\alpha} J_{\nu+\beta-\alpha}(xt)$$

$$\times \left\{ \frac{d}{dx} \int_x^\infty u^{1-\nu}(u^2-x^2)^{\alpha-\beta} \, h(u) \, du \right\} dx.$$

2.1.8 Now, we consider the following class of dual integral equations involving Bessel functions of the first kind of different orders as kernel and arbitrary weight function

$$\int_0^\infty t^{-2\alpha} \, J_\nu(xt)[1+\omega(t)] \, f(t) \, dt \;=\; g(x), \; 0 < x < 1, \tag{2.160}$$

$$\int_0^\infty t^{-2\beta} \, J_\mu(xt) \, f(t) \, dt \;=\; h(x), \; x > 1. \tag{2.161}$$

These are a more general pair than those discussed earlier. Nasim (1986) first considered the above pair and used an operational procedure based on exploiting the properties of the Mellin transform to reduce this pair to a single integral equation involving a Bessel function as kernel. By Hankel inversion this was then reduced to a Fredholm integral equation of the second kind. The method presented by Nasim (1986) is laborious and we do not reproduce it here. However, the general result

44

given by Nasim (1986) was obtained under the assumptions that $|\alpha - \beta| < \frac{\mu - \nu}{2} + 1$ and $\nu \geq 0$, although the second assumption was not clearly stated. This is necessary to make $\lambda > -1$ (where $\lambda = \frac{\mu + \nu}{2} - \alpha + \beta$) for the Hankel inversion to be valid. It is also not clear whether this operational procedure involving the Mellin transform can be used if these restrictions on the parameters μ, ν, α, β are not satisfied. Mandal (1988) re-examined the dual integral equations for arbitrary values of the parameters μ, ν, α and β. He used the *multiplying factor method* of Noble (1963) based on Sonine's integrals to the above class of dual integral equations to reduce it to a Fredholm integral equation of the second kind. The analysis given by Mandal (1988) is the following.

We assume that the parameters μ, ν, α, β appearing in the pair of integral equations (2.160) and (2.161) are most general. By using the formulae

$$\left(\frac{1}{x}\frac{d}{dx}\right)^p [x^{-\mu} J_\mu(xt)] = (-1)^p \, x^{-\mu - p} \, t^p \, J_{\mu + p}(xt), \qquad (2.162)$$

$$\left(\frac{1}{x}\frac{d}{dx}\right)^q [x^{\mu} J_\mu(xt)] = x^{\mu - q} \, t^q \, J_{\mu - q}(xt), \qquad (2.163)$$

where p, q are non-negative integers, it is always possible to change the orders ν, μ to ν', μ' respectively such that $\nu' > -1$, $\mu' > -\frac{3}{2}$ (obviously α, β are then also changed) by choosing p suitably. Thus, without any loss of generality, we can assume that the orders ν, μ of the Bessel functions in (2.160) and (2.161) satisfy the restrictions $\nu > -1$, $\mu > -\frac{3}{2}$. These restrictions are necessary in the analysis that follows. Multiplying each side of the equation (2.160) by $(r^2 - x^2)^\xi \, x^{\nu + 1}$ where $\xi > -1$, integrating from 0 to r (< 1) and using Sonine's first integral in the form

$$\int_0^r (r^2 - x^2)^\xi \, x^{\nu + 1} \, J_\nu(xt) \, dx$$

$$= 2^\xi \, \Gamma(\xi + 1) \, t^{-\xi - 1} \, r^{\xi + \nu + 1} \, J_{\nu + \xi + 1}(rt); \quad \nu > -1, \; \xi > -1, \qquad (2.164)$$

we obtain

$$2^{\xi}\, \Gamma(\xi+1)\, r^{\xi+\nu+1} \int_0^{\infty} t^{-2\alpha-\xi-1}\, J_{\nu+\xi+1}(rt)\, [1+\omega(t)]\, f(t)\, dt$$

$$= \int_0^r x^{\nu+1}\, (r^2 - x^2)^{\xi}\, g(x)\, dx, \ 0 < r < 1. \tag{2.165}$$

Similarly, multiplying each side of equation (2.161) by $(x^2 - r^2)^{\eta}\, x^{-\mu+1}$ where $\eta > -1$, integrating from $r\ (> 1)$ to ∞ and using Sonine's second integral in the form

$$\int_r^{\infty} (x^2 - r^2)^{\eta}\, x^{-\mu+1}\, J_{\mu}(xt)\, dx$$

$$= 2^{\eta}\, \Gamma(\eta+1)\, t^{-\eta-1}\, r^{-\mu+\eta+1}\, J_{\mu-\eta-1}(rt), \ -1 < \eta < \frac{\mu}{2} - \frac{1}{4}, \tag{2.166}$$

we obtain

$$2^{\eta}\, \Gamma(1+\eta)\, r^{-\mu+\eta+1} \int_0^{\infty} t^{-2\beta-\eta-1}\, J_{\mu-\eta-1}(rt)\, f(t)\, dt$$

$$= \int_r^{\infty} (x^2 - r^2)^{\eta}\, x^{-\mu+1}\, h(x)\, dx, \ 1 < r < \infty. \tag{2.167}$$

Using the relation (2.163) (with q replaced by l and x replaced by r) in (2.165), we find that

$$\int_0^{\infty} t^{-2\alpha-\xi-1+l}\, J_{\nu+\xi+1-l}\,(rt)\, [1+\omega(t)]\, f(t)\, dt$$

$$= \frac{2^{-\xi}}{\Gamma(\xi+1)} r^{-\xi-\nu-1+l} \left(\frac{1}{r}\frac{d}{dr}\right)^l \int_0^r x^{\nu+1}(r^2 - x^2)^{\xi}\, g(x)\, dx, \ 0 < r < 1, \tag{2.168}$$

where l is a non-negative integer. Similarly, using the relation (2.162) (with p replaced by m and x replaced by r) in (2.167), we find that

$$\int_0^{\infty} t^{-2\beta-\eta-1+m}\, J_{\mu-\eta-1+m}(rt)\, f(t)\, dt$$

$$= \frac{(-1)^m\, 2^{-\eta}}{\Gamma(1+\eta)}\, r^{\mu-\eta-1+m} \left(\frac{1}{r}\frac{d}{dr}\right)^m \int_r^{\infty} (x^2 - r^2)^{\eta}\, x^{-\mu+1}\, h(x)\, dx, \ r > 1, \tag{2.169}$$

where m is a non-negative integer.

46

In equations (2.168) and (2.169), we now equate the powers of t and the orders of the Bessel functions. This gives two equations to determine ξ, η as

$$-2\alpha - \xi - 1 + l = -2\beta - \eta - 1 + m,$$
$$\nu + \xi + 1 - l = \mu - \eta - 1 + m.$$

Solving these two equations, we get

$$\left.\begin{aligned} \xi &= \frac{\mu - \nu}{2} + \beta - \alpha - 1 + l, \\ \eta &= \frac{\mu - \nu}{2} + \alpha - \beta - 1 + m. \end{aligned}\right\} \tag{2.170}$$

The requirements that ξ and η must be greater than -1 can be satisfied by choosing the non-negative integers l and m appropriately. Hence, we obtain a single integral equation

$$\int_0^\infty t^{-\gamma} J_\lambda(rt) \, f(t) \, dt$$

$$= \begin{cases} G(r) - \int_0^\infty t^{-\gamma} J_\lambda(rt) \, \omega(t) \, f(t) \, dt, \ 0 < r < 1, \\ H(r), \ r > 1, \end{cases} \tag{2.171}$$

where $\lambda = \dfrac{\mu + \nu}{2} + \beta - \alpha$, $\gamma = \dfrac{\mu - \nu}{2} + \alpha + \beta$, and $G(r), H(r)$ are given by the right sides in equations (2.168) and (2.169) respectively.

Now in the single integral equation (2.171), we require $\lambda > -1$ for Hankel inversion. In general, this requirement may not be satisfied. To overcome this difficulty, we use the relation (2.162) (with p replaced by s and x replaced by r) in (2.171) to obtain

$$\int_0^\infty t^{-\gamma+s} J_{\lambda+s}(rt) \, f(t) \, dt$$

$$= \begin{cases} G_1(r) - (-1)^s \int_0^\infty t^{-\gamma+s} \, J_{\lambda+s}(rt) \, \omega(t) \, f(t) \, dt, \ 0 < r < 1, \\ H_1(r), \ r > 1, \end{cases} \tag{2.172}$$

where

$$\left.\begin{aligned} G_1(r) &= (-1)^s \, r^{\lambda+s} \left(\frac{1}{r}\frac{d}{dr}\right)^s \left[r^{-\lambda} \, G(r)\right], \\ H_1(r) &= (-1)^s \, r^{\lambda+s} \left(\frac{1}{r}\frac{d}{dr}\right)^s \left[r^{-\lambda} \, H(r)\right], \end{aligned}\right\} \tag{2.173}$$

and s is a non-negative integer. We can always choose $\lambda + s > -1$ whatever α, β may be. Then the Hankel inversion reduces equation (2.172) to produce a Fredholm integral equation of the second kind for the function $f(x)$ as

$$f(x) = x^{\gamma - s + 1}\left[\int_0^1 r\, G_1(r)\, J_{\lambda + s}(rx)\, dr + \int_1^\infty r\, H_1(r)\, J_{\lambda + s}(rx)\, dr\right.$$

$$\left. - \int_0^\infty u^{-\gamma + s}\omega(u)\, \frac{\{uJ_{\lambda + s + 1}(u)J_{\lambda + s}(x) - xJ_{\lambda + s + 1}(x)J_{\lambda + s}(u)\}}{u^2 - x^2} f(u)\, du\right],$$

(2.174)

where λ, γ are given by (2.171) and s is a non-negative integer to be chosen appropriately so as to make $\lambda + s > -1$.

Special cases

(i) $\mu = \nu$, $\alpha = -\frac{a}{2}$, $\beta = 0$ where $\nu > \max(-1, \frac{1}{2} - a)$ and $0 < a < 2$. Then the dual integral equations are reduced to

$$\int_0^\infty t^a\, [1 + \omega(t)]\, J_\nu(xt)\, f(t)]\, dt = g(x),\quad 0 < x < 1, \tag{2.175}$$

$$\int_0^\infty J_\nu(xt)\, f(t)\, dt = h(x),\quad x > 1. \tag{2.176}$$

These dual integral equations were considered by Noble (1963). The conditions $\nu > -1$ and $\mu > -\frac{3}{2}$ are automatically satisfied. Now from (2.170),

$$\xi = -1 + \frac{a}{2} + l,\quad \eta = -1 - \frac{a}{2} + m,$$

where l, m are non-negative integers to be chosen such that $\xi > -1$, $\eta > -1$. We choose $l = 0$, $m = 1$ so that $\xi = -1 + \frac{a}{2} > -1$ and $\eta = -\frac{a}{2} > -1$. Also $\lambda = \nu + \frac{a}{2} > -1$, $\gamma = -\frac{a}{2}$. Hence, we take $s = 0$ in the relation (2.174). Then

$$G_1(r) = G(r) = \frac{2^{1 - a/2}}{\Gamma(a/2)}\, r^{-\nu - \frac{a}{2}}\int_0^r u^{\nu + 1}(r^2 - u^2)^{-1 + \frac{a}{2}}g(u)\, du,$$

$$H_1(r) = H(r) = \frac{-2^{a/2}}{\Gamma(1 - a/2)}\, r^{-\nu + \frac{a}{2} - 1}\frac{d}{dr}\int_r^\infty (u^2 - r^2)^{-\frac{a}{2}}\, u^{1 - \nu}\, h(u)\, du,$$

48

so that the Fredholm integral equation for $f(x)$ is

$$f(x) = \frac{(2x)^{1-\frac{a}{2}}}{\Gamma(a/2)} \int_0^1 r^{1-\nu-\frac{a}{2}} \left\{ \int_0^r u^{1+\nu} (r^2-u^2)^{-1+\frac{a}{2}} g(u) \, du \right\} J_{\nu+\frac{a}{2}}(rx) \, dr$$

$$- \frac{x^{1-\frac{a}{2}} 2^{a/2}}{\Gamma(1-a/2)} \int_1^\infty r^{-\nu+\frac{a}{2}} \frac{d}{dr} \left\{ \int_r^\infty (u^2-r^2)^{-\frac{a}{2}} u^{1-\nu} h(u) \, du \right\} J_{\nu+\frac{a}{2}}(rx) \, dr$$

$$- x^{1-\frac{a}{2}} \int_0^\infty \omega(u) \frac{\left\{ u \, J_{\nu+\frac{a}{2}+1}(u) \, J_{\nu+\frac{a}{2}}(x) - x J_{\nu+\frac{a}{2}+1}(x) J_{\nu+\frac{a}{2}}(x) \right\}}{(u^2-x^2) \, u^{-\frac{a}{2}}} f(u) \, du.$$

This result was obtained by Noble (1963) (cf. equation (4.7) there). Putting $h(x) = 0$, $w(u) = 0$, we obtain the closed form solution of the dual integral equations

$$\int_0^\infty t^a f(t) J_\nu(xt) \, dt = g(x), \quad 0 < x < 1,$$

$$\int_0^\infty f(t) J_\nu(xt) \, dt = 0, \quad x > 1,$$

where $\nu > \max(-1, \frac{1}{2} - a)$, $0 < a < 2$, as given by

$$f(x) = \frac{(2x)^{1-\frac{a}{2}}}{\Gamma(\frac{a}{2})} \int_0^1 r^{\frac{a}{2}+1} J_{\nu+\frac{a}{2}}(rx) \left\{ \int_0^1 s^{\nu+1} (1-s^2)^{\frac{a}{2}-1} g(rs) \, ds \right\} dr.$$

This coincides with the result obtained by Titchmarsh (1937).

(ii) $|\alpha - \beta| < \dfrac{\mu - \nu}{2} + 1$ and $\nu \geq 0$.

This case was considered by Nasim (1986). Here $\lambda > -1$, and we may mention here that the restriction $\nu \geq 0$ was not stated explicitly by Nasim (1986). Unless this restriction holds good, it is not possible to use Hankel inversion with the order of the Bessel function as λ. To make $\xi > -1$, $\eta > -1$, we choose $l = 1$, $m = 1$ in (2.170). Also as $\lambda > -1$, we choose $s = 0$ in (2.174) so that the Fredholm integral equation (2.174) in this case reduces to

$$f(x) = x^{\gamma+1} \left[\frac{x^{-\gamma+2\alpha}}{\Gamma(\gamma-2\alpha+1)} \int_0^1 r^{-\gamma+2\alpha-\nu} J_\lambda(rx) \, d \left\{ \int_0^r u^{\nu+1}(r^2-u^2)^{\gamma-2\alpha} g(u) \, du \right\} \right.$$

$$- \frac{x^{-\gamma+2\beta}}{\Gamma(\gamma-2\beta+1)} \int_1^\infty r^\lambda J_\lambda(rx) \, d \left\{ \int_r^\infty (u^2-r^2)^{\gamma-2\beta} u^{1-\mu} h(u) \, du \right\}$$

$$\left. - \int_0^\infty u^{-\gamma} \omega(u) \frac{\{u J_{\lambda+1}(u) J_\lambda(x) - x J_{\lambda+1}(x) J_\lambda(u)\}}{u^2 - x^2} f(u) \, du \right]. \tag{2.177}$$

49

This result does not seem to coincide with the result obtained by Nasim (1986). However, the known results for the following two particular cases given in the book by Sneddon (1966) as well as in the paper by Nasim (1986) are deduced from the above result (2.177) implying that this result is also correct.

When $\mu = \nu = 0$, $\alpha = \frac{1}{2}$, $\beta = 0$, (2.170) gives $\lambda = -\frac{1}{2}$ and $\gamma = \frac{1}{2}$. Substituting in (2.177), after some simplification, we obtain

$$f(x) = \frac{2x}{\pi} \int_0^1 \cos rx \, d\left\{ \int_0^r \frac{u \, g(u)}{\sqrt{r^2 - u^2}} \, du \right\}$$

$$+ \frac{2x}{\pi} \int_1^\infty \cos rx \left\{ \int_r^\infty \frac{u \, h(u)}{\sqrt{u^2 - r^2}} \, du \right\} dr$$

$$- \frac{x}{\pi} \int_0^\infty \frac{\omega(u)}{u} \left\{ \frac{\sin(u + x)}{u + x} + \frac{\sin(u - x)}{u - x} \right\} f(u) \, du. \qquad (2.178)$$

This result can also be deduced directly from (2.174) by choosing $s = 0$, $l = 1$, $m = 0$.

When $\mu = \nu = 0$, $\alpha = -\frac{1}{2}$, $\beta = 0$, the integral equation for $f(x)$ can similarly be deduced from (2.177). After some simplification, we get

$$f(x) = \frac{2}{\pi} \int_0^1 \sin rx \left\{ \int_0^r \frac{u \, g(u)}{\sqrt{r^2 - u^2}} \, du \right\} dr$$

$$- \frac{2}{\pi} \int_1^\infty \sin rx \left\{ \int_r^\infty \frac{u \, h(u)}{\sqrt{u^2 - r^2}} \, du \right\} dr$$

$$- \frac{1}{\pi} \int_0^\infty \omega(u) \left\{ \frac{\sin(u - x)}{u - x} - \frac{\sin(u + x)}{u + x} \right\} f(u) \, du.$$

This can also be deduced directly from (2.174) by choosing $s = 0$, $l = 0$, $m = 1$.

2.1.9 Next, we consider the pair (cf. Rahman 1995)

$$\int_0^\infty t^{-2\alpha} \left[1 + \omega(t) \right] f(t) \, J_\nu(xt) \, dt = G(x), \quad 0 < x < 1, \qquad (2.179)$$

$$\int_0^\infty f(t) \, J_\nu(xt) \, dt = H(x), \quad 1 < x < \infty, \qquad (2.180)$$

50

where $\omega(t), G(x)$ and $H(x)$ are known functions with $\omega(t) \to 0$ as $t \to \infty$. Putting $f(t) = t\,\psi(t)$, $g(x) = (\frac{2}{x})^{2\alpha}\, G(x)$, $h(x) = H(x)$, the above pair can be rewritten as

$$S_{\frac{\nu}{2}-\alpha,\ 2\alpha}[1 + \omega(x)]\,\psi(x) = g(x),\ 0 < x < 1, \tag{2.181}$$

$$S_{\frac{\nu}{2},\ 0}\,\psi(x) = h(x),\ 1 < x < \infty, \tag{2.182}$$

where $S_{\beta,\ \alpha}\,\psi(x)$ is the Hankel operator defined by

$$S_{\beta,\ \alpha}\,\psi(x) = 2^{\alpha} x^{-\alpha} \int_0^\infty t^{1-\alpha}\, J_{\alpha+2\beta}(xt)\,\psi(t)\,dt. \tag{2.183}$$

In equations (2.181) and (2.182), we have

$$g(x) = \begin{cases} g_1(x),\ 0 < x < 1, \\ g_2(x),\ 1 < x < \infty, \end{cases}$$

and

$$h(x) = \begin{cases} h_1(x),\ 0 < x < 1, \\ h_2(x),\ 1 < x < \infty, \end{cases}$$

where $g_1(x)$ and $h_2(x)$ are known but $g_2(x)$ and $h_1(x)$ are unknown functions.

The dual integral equation pair (2.179), (2.180) or (2.181), (2.182) with $h_2(x) = 0$ arises in many mixed boundary value problems in the theory of elasticity. In particular, the dynamic problem of a penny-shaped crack embedded in an elastic solid or the dynamic problem of a circular punch resting on an elastic half-space can be reduced to the dual integral equation pair (2.179), (2.180) or (2.181), (2.182). In all these problems, the unknown function $h_1(x)$ represents a quantity of considerable physical interest. For example, in crack problems this function is related to the crack opening displacement, while in contact problems it defines the unknown contact pressure under the punch. These quantities are important in linear fracture and contact mechanics, since a knowledge of these quantities allows one to determine other physical quantities of interest, namely, stress intensity factor, displacement under the punch, etc. To solve these pairs, Rahman (1995) gave a method which is particularly suitable for the efficient numerical solution of mixed boundary value

51

problems of elasticity, heat transfer, acoustics, electrostatics and electrodynamics. In this method, the pair of integral equations is first reduced to a Fredholm integral equation of the second kind through an unknown auxiliary function. Then this integral equation is reduced into an infinite system of linear algebraic equations by representing the unknown auxiliary function in the form of an infinite series of Jacobi polynomials. We describe this method here briefly.

To reduce the dual integral equations (2.179) and (2.180) to a Fredholm integral equation of the second kind, we put

$$\psi(x) = S_{\frac{\nu}{2}, \, -\alpha} \, \phi(x)$$

into equations (2.181) and (2.182), so that we get

$$I_{\frac{\nu}{2}, \, \alpha} \, \phi(x) + S_{\frac{\nu}{2}-\alpha, \, 2\alpha} \, \omega(x) \, S_{\frac{\nu}{2}, \, -\alpha} \, \phi(x) \quad = \quad g(x), \qquad (2.184)$$

$$K_{\frac{\nu}{2}, \, -\alpha} \, \phi(x) \quad = \quad h(x), \qquad (2.185)$$

where $I_{\beta,\alpha}$ and $K_{\beta,\alpha}$ are the Erdélyi–Kober operators (cf. Sneddon 1966, p. 48) defined by

$$
\left. \begin{aligned}
I_{\beta, \, \alpha} \, f(x) &= \frac{2x^{-2\alpha-2\beta}}{\Gamma(\alpha)} \int_0^x (x^2 - u^2)^{\alpha-1} \, u^{2\beta+1} \, f(u) \, du, \; 0 < \alpha < 1, \\
&= \frac{x^{-2\alpha-2\beta-1}}{\Gamma(1+\alpha)} \frac{d}{dx} \int_0^x (x^2 - u^2)^{\alpha} \, u^{2\beta+1} \, f(u) \, du, \; -1 < \alpha < 0,
\end{aligned} \right\}
$$
$$(2.186)$$

$$
\left. \begin{aligned}
K_{\beta, \, \alpha} \, f(x) &= \frac{2x^{2\beta}}{\Gamma(\alpha)} \int_x^{\infty} (u^2 - x^2)^{\alpha-1} \, u^{-2\alpha-2\beta+1} \, f(u) \, du, \; 0 < \alpha < 1, \\
&= \frac{x^{2\beta-1}}{\Gamma(1+\alpha)} \frac{d}{dx} \int_x^{\infty} (u^2 - x^2)^{\alpha} \, u^{-2\alpha-2\beta+1} \, f(u) \, du, \; -1 < \alpha < 0,
\end{aligned} \right\}
$$
$$(2.187)$$

where $\Gamma(\alpha)$ is the gamma function.

In deriving equations (2.184) and (2.185), we have also used the following two relations (cf. Sneddon 1966, p. 50).

$$\left.\begin{array}{rcl} S_{\lambda+\alpha,\,\beta}\, S_{\lambda,\,\alpha} &=& I_{\lambda,\,\alpha+\beta}, \\[4pt] S_{\lambda,\,\alpha}\, S_{\lambda+\alpha,\,\beta} &=& K_{\lambda,\,\alpha+\beta}. \end{array}\right\} \tag{2.188}$$

Solving the fractional equations (2.184) and (2.185), we obtain

$$\phi(x) = I_{\frac{\nu}{2},\,\alpha}^{-1}\, g(x) - S_{\frac{\nu}{2}-\alpha,\,\alpha}\, \omega(x)\, S_{\frac{\nu}{2},\,-\alpha}\, \phi(x), \tag{2.189}$$

$$\phi(x) = K_{\frac{\nu}{2},\,-\alpha}^{-1}\, h(x), \tag{2.190}$$

where I^{-1} and K^{-1} are the inverse operators defined by

$$\left.\begin{array}{rcl} I_{\beta,\,\alpha}^{-1} &=& I_{\beta+\alpha,\,-\alpha}, \\[4pt] K_{\beta,\,\alpha}^{-1} &=& K_{\beta+\alpha,\,-\alpha}. \end{array}\right\} \tag{2.191}$$

Now we write equations (2.189) and (2.190) in the subintervals $[0,1]$ and $[1,\infty)$ as given by

$$\phi_1(x) + E(x) = \binom{x}{0} I_{\frac{\nu}{2},\,\alpha}^{-1}\, g_1(x),\ 0 \le x \le 1, \tag{2.192}$$

$$\phi_2(x) + E(x) = \binom{1}{0} I_{\frac{\nu}{2},\,\alpha}^{-1}\, g_1(x) + \binom{x}{1} I_{\frac{\nu}{2},\,\alpha}^{-1}\, g_2(x),\ 1 \le x < \infty, \tag{2.193}$$

$$\phi_2(x) = \binom{\infty}{x} K_{\frac{\nu}{2},\,-\alpha}^{-1}\, h_2(x),\ 1 \le x < \infty, \tag{2.194}$$

$$\phi_1(x) = \binom{\infty}{1} K_{\frac{\nu}{2},\,-\alpha}^{-1}\, h_2(x) + \binom{1}{x} K_{\frac{\nu}{2},\,-\alpha}^{-1}\, h_1(x),\ 0 \le x \le 1, \tag{2.195}$$

where

$$E(x) = S_{\frac{\nu}{2}-\alpha,\,\alpha}\, \omega(x)\, S_{\frac{\nu}{2},\,-\alpha}\, \phi(x) = x^{-\alpha} \int_0^\infty u^{1+\alpha} K(x,u)\, \phi(u)\, du, \tag{2.196}$$

$$K(x,u) = \int_0^\infty t\, \omega(t)\, J_{\nu-\alpha}(xt)\, J_{\nu-\alpha}(ut)\, dt. \tag{2.197}$$

In the above equations, $\phi_1(x)$ and $\phi_2(x)$ denote the two parts of the function $\phi(x)$ corresponding to the intervals $[0,1]$ and $[1,\infty)$. The quantities in parentheses indicate the new limits of integration.

We consider the special case for which $h_2(x) = 0$. From (2.194), we find that $\phi_2(x) = 0$ and from equation (2.192), we obtain the Fredholm integral equation of the second kind as given by

$$\phi_1(x) + x^{-\alpha} \int_0^1 u^{1+\alpha} K(x,u)\, \phi_1(u)\, du = p(x), \tag{2.198}$$

where

$$p(x) = I_{\frac{\nu}{2}, \alpha}^{-1} \, g_1(x) = I_{\frac{\nu}{2}+\alpha, -\alpha} \, g_1(x). \tag{2.199}$$

Let us assume that the solution of the integral equation (2.198) is of the form

$$\phi_1(x) = x^{\nu-2\alpha} \sum_{n=0}^{\infty} a_n \, P_n^{\nu-\alpha, \, 0} \, (1 - 2x^2), \tag{2.200}$$

where $P_n^{m, \, l}(x)$'s are Jacobi polynomials and a_n's are the unknown constants to be determined. Substituting (2.200) into (2.198), then using the integral representation (A.1.26) and the relation (cf. Rahman 1995)

$$\int_0^1 \frac{P_n^{a, \, b}(1 - 2x^2) \, P_m^{a, \, b}(1 - 2x^2)}{2^{-2-a-b} \, x^{-1-2a}(1 - x^2)^{-b}} \, dx$$

$$= \frac{2^{a+b+1}\Gamma(a+n+1)\Gamma(b+n+1)}{n! \, \Gamma \, (n+a+b+1) \, (\alpha+\beta+2n+1)} \, \delta_{mn}, \tag{2.201}$$

where δ_{mn} is the Kronecker delta, we obtain an infinite series of linear algebraic equations as given by

$$\frac{a_m}{2(1+\nu-\alpha+2m)} + \sum_{n=0}^{\infty} a_n \, W_{mn} = p_m, \tag{2.202}$$

where

$$W_{mn} = \int_0^\infty \frac{\omega(t)}{t} J_{1+\nu-\alpha+2n}(t) \, J_{1+\nu-\alpha+2m}(t) \, dt, \tag{2.203}$$

$$p_m = \int_0^1 x^{1+\nu} \, p(x) \, P_m^{\nu-\alpha, \, 0} \, (1 - 2x^2) \, dx. \tag{2.204}$$

The infinite series of linear algebraic equations (2.202) can be solved numerically.

An Application

We consider the axisymmetric problem of torsion of a rigid circular disc embedded at a depth h in a homogeneous, isotropic elastic half-space. The bonded disc is assumed to be rotated about the z-axis through an angle θ.

The mathematical formulation of the problem is

$$\frac{\partial^2 u}{\partial r^2} + \frac{1}{r}\frac{\partial u}{\partial r} + \frac{\partial^2 u}{\partial z^2} - \frac{u}{r^2} = 0, \tag{2.205}$$

54

with the boundary conditions

$$\left.\begin{array}{rcl}
\sigma_{\theta z}(r,h) & = & 0,\ 0 \leq r < \infty, \\
\sigma_{\theta z}(r,0) & = & 0,\ r > 1, \\
u(r,0) & = & \theta r,\ 0 < r < 1.
\end{array}\right\} \tag{2.206}$$

All dimensions are normalized with respect to the radius of the disc. The above mixed boundary value problem is reduced to the solution of the dual integral equations

$$\int_0^\infty t^{-1}(1 + e^{-2ht})\ f(t)\ J_1(rt)\ dt\ =\ 2\theta\ \mu r,\ 0 < r < 1, \tag{2.207}$$

$$\int_0^\infty f(t)\ J_1(rt)\ dt\ =\ 0,\ r > 1, \tag{2.208}$$

where μ is the shear modulus of the embedding medium and

$$\sigma_{\theta z}(r,0) = \int_0^\infty f(t)\ J_1(rt)\ dt. \tag{2.209}$$

Following (2.198), the dual integral equations (2.207) and (2.208) are reduced to the Fredholm integral equation of the second kind

$$\phi_1(r) + r^{-\frac{1}{2}} \int_0^1 u^{3/2}\ K(r,u)\ \phi_1(u)\ du = p(r) \tag{2.210}$$

where

$$K(r,u)\ =\ \int_0^\infty t\ e^{-2ht}\ J_{\frac{1}{2}}(rt)\ J_{\frac{1}{2}}(ut)\ dt, \tag{2.211}$$

$$p(r)\ =\ I_{\frac{1}{2},\frac{1}{2}}^{-1}\ g_1(r),\ \ g_1(r) = 4\theta\mu. \tag{2.212}$$

Equation (2.200) suggests that we can express the solution of (2.210) as

$$\phi_1(r) = \sum_{n=0}^\infty a_n\ P_n^{\frac{1}{2},\ 0}(1 - 2r^2). \tag{2.213}$$

Then the corresponding infinite system of linear algebraic equations is

$$\frac{a_m}{4m + 3} + \sum_{n=0}^\infty a_n\ W_{mn} = p_m, \tag{2.214}$$

55

where

$$W_{mn} = \int_0^\infty t^{-1} e^{-2ht} J_{3/2+2n}(t) \ J_{3/2+2m}(t) \ dt, \qquad (2.215)$$

$$p_m = \int_0^1 r^2 \ p(r) \ P_m^{\frac{1}{2}, \ 0}(1 - 2r^2) \ dr. \qquad (2.216)$$

Using the relation

$$P_m^{\frac{1}{2}, \ 0}(1 - 2r^2) = (-1)^m \ r^{-1} \ P_{2m+1}(r), \qquad (2.217)$$

$P_{2m+1}(r)$ being Legendre's polynomial, we find from (2.216) that

$$p_m = 0, \ m > 0 \ \text{ and } p_0 = \frac{8 \ \theta \mu}{3\sqrt{\pi}}. \qquad (2.218)$$

Putting $b_m = 3a_m\sqrt{\pi}/8\theta\mu$, the system (2.214) becomes

$$\frac{b_m}{4m + 3} + \sum_{n=0}^\infty b_m \ W_{mn} = \delta_{m0}. \qquad (2.219)$$

The corresponding singular stress $\sigma_{\theta z}(r, 0)$ beneath the disc is given by

$$\sigma_{\theta z}(r, 0) = r \ p_0 \sum_{n=0}^\infty b_n \ \frac{n!}{\Gamma(n + \frac{1}{2})} \ (1 - r^2)^{-\frac{1}{2}} \ P_n^{1, -\frac{1}{2}}(1 - 2r^2). \qquad (2.220)$$

Then, the stress intensity factor is given by

$$K = \lim_{r \to 1-} \sqrt{2\pi(1 - r)} \ \sigma_{\theta z}(r, 0). \qquad (2.221)$$

Substituting (2.220) into equation (2.221), we get

$$K = p_0 \sum_{n=0}^\infty (-1)^n \ b_n, \qquad (2.222)$$

where we have used the relation

$$P_n^{a, \ b}(-1) = (-1)^n \frac{\Gamma(n + b + 1)}{\Gamma(n + 1)\Gamma(1 + b)}.$$

For arbitrary h, the infinite system (2.219) can be solved numerically to determine the stress intensity factor.

2.2 Kernels involving a Bessel function of the second kind

In this section, we present some dual integral equations involving Bessel functions of the second kind considered by Nasim and Aggarwala (1984). First we define the following operators $I_{\eta,\alpha}(a, x : 2)$ and $K_{\eta,\alpha}(x, b : 2)$ which are simple generalizations of Erdélyi–Kober operators (cf. Nasim and Aggarwala 1984):

$$
I_{\eta,\,\alpha}(a, x : 2)[f] =
\begin{cases}
\dfrac{2}{\Gamma(\alpha)} x^{-2(\eta+\alpha)} \displaystyle\int_a^x (x^2 - t^2)^{\alpha-1}\, t^{2\eta+1} f(t)\, dt, \; \alpha > 0, \\[2ex]
\dfrac{x^{-(2\eta+2\alpha+1)}}{\Gamma(1+\alpha)} \dfrac{d}{dx} \displaystyle\int_a^x (x^2 - t^2)^{\alpha-1}\, t^{2\eta+1}\, f(t)\, dt, \; -1 < \alpha < 0,
\end{cases}
$$

and

$$
K_{\eta,\,\alpha}(x, b : 2)[f] =
\begin{cases}
\dfrac{2}{\Gamma(\alpha)} x^{2\eta} \displaystyle\int_x^b (t^2 - x^2)^{\alpha-1}\, t^{1-2\alpha-2\eta}\, f(t)\, dt, \; \alpha > 0, \\[2ex]
-\dfrac{x^{2\eta-1}}{\Gamma(1+\alpha)} \dfrac{d}{dx} \displaystyle\int_x^b (t^2 - x^2)^{\alpha}\, t^{1-2\alpha-2\eta}\, f(t)\, dt, \; -1 < \alpha < 0.
\end{cases}
$$

We also use the notation

$$
I_{\eta,\,\alpha}(0, x : 2) = I_{\eta,\,\alpha} \quad \text{and} \quad K_{\eta,\,\alpha}(x, \infty : 2) = K_{\eta,\,\alpha}.
$$

Defining the inverse operators, it can be shown that

$$
\begin{aligned}
I_{\eta,\,\alpha}^{-1}(a, x : 2) &= I_{\eta+\alpha,-\alpha}(a, x : 2), \\
K_{\eta,\,\alpha}^{-1}(x, b : 2) &= K_{\nu+\alpha,-\alpha}(x, b : 2).
\end{aligned}
$$

Now, we consider the dual integral equations

$$
\int_0^\infty t^{-2\alpha}\, f(t)\, Y_\mu(xt)\, dt = g(x),\; 0 < x < 1, \tag{2.223}
$$

$$
\int_0^\infty t^{-2\beta}\, f(t)\, Y_\nu(xt)\, dt = h(x),\; 1 < x < \infty, \tag{2.224}
$$

where $2(\alpha - \beta) = \nu - \mu$, $|\mu| < \frac{1}{2}$ and $|\nu| < \frac{1}{2}$, $Y_\mu(z)$ being the Bessel function of the second kind.

The solution of the above pair is obtained for the following two cases:

Case I: $h(x) = 0$

The above dual integral equations become

$$\int_0^\infty t^{-2\alpha} \; f(t) \; Y_\mu(xt) \; dt \;=\; g(x), \; 0 < x < 1, \tag{2.225}$$

$$\int_0^\infty t^{-2\beta} \; f(t) \; Y_\nu(xt) \; dt \;=\; 0, \; 1 < x < \infty. \tag{2.226}$$

We assume that

$$\int_0^\infty t^{-2\beta} \; f(t) \; Y_\nu(xt) \; dt = \phi(x) \; H(1-x), \tag{2.227}$$

where $\phi(x)$ is an unknown function, $H(x)$ being the Heaviside unit function. From equations (2.226) and (2.227), by the inverse Y-transform theorem, we get

$$t^{-2\beta-\frac{1}{2}} f(t) = \int_0^1 \phi(u) \; H_\nu(ut) \; (ut)^{\frac{1}{2}} du, \; |\nu| < \frac{1}{2}, \tag{2.228}$$

where $H_\nu(x)$ is Struve function of order ν. Now substituting the value of $f(t)$ from (2.228) into equation (2.225) and then simplifying, we get

$$g(x) = \int_0^1 \sqrt{u} \; \phi(u) \left\{ \int_0^\infty t^{\mu-\nu+1} \; Y_\mu(xt) \; H_\nu(ut) \; dt \right\} \; du, \; 0 < x < 1,$$

where $2(\alpha - \beta) = \nu - \mu$. Using a result in Erdélyi et al. (1954b), p. 114, the above expression is simplified to the form

$$g(x) = \frac{2^{\mu-\nu+1}}{\Gamma(\nu-\mu)} \; x^\mu \int_x^1 u^{\frac{1}{2}-\nu} (u^2 - x^2)^{\nu-\mu-1} \phi(u) \; du, \; 0 < x < 1,$$

where $\nu - \mu > 0$. The above relation can be represented by

$$g(x) = 2^{\mu-\nu} \; K_{\frac{\mu}{2}, \; \nu-\mu}(x, 1 : 2) \; \left[u^{\nu-\mu-\frac{1}{2}} \phi(u) \right],$$

which gives

$$\phi(x) = 2^{\nu-\mu} \; x^{\mu-\nu+\frac{1}{2}} \; K_{\nu-\frac{\mu}{2}, \; \mu-\nu}(x, 1 : 2)[g].$$

Substituting this $\phi(x)$ into equation (2.228), the solution of the pair (2.225) and (2.226) is given by

$$f(t) = 2^{\nu-\mu} \; t^{2\beta+1} \int_0^1 u^{\mu-\nu+1} \; H_\nu(ut) \; K_{\nu-\frac{\mu}{2}, \; \mu-\nu}(u, 1 : 2)[g] \; du. \tag{2.229}$$

If $\nu - \mu < 0$, then the operator K is to be defined accordingly.

Case II: $g(x) = 0$

In this case, the dual integral equations are

$$\int_0^\infty t^{-2\alpha} \, f(t) \, Y_\mu(xt) \, dt \;=\; 0, \; 0 < x < 1, \tag{2.230}$$

$$\int_0^\infty t^{-2\beta} \, f(t) \, Y_\nu(xt) \, dt \;=\; h(x), \; 1 < x < \infty. \tag{2.231}$$

As in the previous case, by the inverse Y-transform theorem, for an appropriately defined unknown function $\phi(x)$, the equation (2.230) gives

$$f(t) = t^{2\alpha+1} \int_1^\infty \phi(u) \, H_\mu(ut) \, \sqrt{u} \, du, \; |\mu| < \frac{1}{2}. \tag{2.232}$$

Substituting the above $f(t)$ into (2.231) and simplifying, we obtain

$$h(x) \;=\; \int_1^\infty \sqrt{u} \, \phi(u) \left\{ \int_0^\infty t^{\nu-\mu+1} \, Y_\nu(xt) \, H_\mu(ut) \, dt \right\} du$$

$$\;=\; \frac{2^{\nu-\mu+1}}{\Gamma(\mu-\nu)} \, x^\nu \int_x^\infty u^{\frac{1}{2}-\mu}(u^2 - x^2)^{\mu-\nu-1} \phi(u) \, du, \; \mu - \nu > 0$$

$$\;=\; 2^{\nu-\mu} \, K_{\frac{\nu}{2}, \, \mu-\nu} \left[u^{\mu-\nu-\frac{1}{2}} \phi(u) \right], \; 1 < x < \infty,$$

which produces

$$\phi(x) = 2^{\mu-\nu} x^{\nu-\mu+\frac{1}{2}} \, K_{\mu-\frac{\nu}{2}, \, \nu-\mu}[h]. \tag{2.233}$$

Using the relation (2.233) in (2.232), the solution of the dual integral equations (2.230) and (2.231) is given by

$$f(t) = 2^{\mu-\nu} \, t^{2\alpha+1} \int_1^\infty u^{\nu-\mu+1} \, H_\mu(ut) \, K_{\mu-\frac{\nu}{2}, \, \nu-\mu}[h] \, du. \tag{2.234}$$

This is also the solution even if $\mu - \nu < 0$.

Combining the results (2.229) and (2.234), the solution of the pair (2.223) and (2.224) is obtained as

$$f(t) \;=\; 2^{\nu-\mu} \, t^{2\beta+1} \int_0^1 u^{\mu-\nu+1} \, H_\nu(ut) \, K_{\nu-\frac{\mu}{2}, \, \mu-\nu}(u, 1:2)[g] \, du$$

$$+ 2^{\mu-\nu} \, t^{2\alpha+1} \int_1^\infty u^{\nu-\mu+1} \, H_\mu(ut) \, K_{\mu-\frac{\nu}{2}, \, \nu-\mu}[h] \, du. \tag{2.235}$$

59

In particular, if $0 < \nu - \mu < 1$, then the above relation (2.235) becomes

$$f(t) = -\frac{2^{\nu-\mu}}{\Gamma(1+\mu-\nu)} \, t^{2\beta+1} \int_0^1 u^\nu H_\nu(ut) \left\{ \frac{d}{du} \int_u^1 (x^2 - u^2)^{\mu-\nu} x^{1-\mu} g(x) dx \right\} \, du$$

$$+ \frac{2^{\mu-\nu}}{\Gamma(\mu-\nu)} \int_1^\infty u^{1+\mu} H_\mu(ut) \left\{ \int_u^\infty (x^2 - u^2)^{\nu-\mu-1} x^{1-\nu} \, h(x) \, dx \right\} du.$$

If $-1 < \nu - \mu < 0$, then

$$f(t) = \frac{2^{1-\nu-\mu}}{\Gamma(\mu-\nu)} \, t^{2\beta+1} \int_0^1 u^{1+\nu} H_\nu(ut) \left\{ \int_u^1 x^{1-\mu} (x^2 - u^2)^{\mu-\nu-1} \, g(x) \, dx \right\} \, du$$

$$+ \frac{2^{\mu-\nu} t^{2\alpha+1}}{\Gamma(1+\nu-\mu)} \int_1^\infty u^\mu H_\mu(ut) \left\{ \int_u^\infty x^{1-\nu} (x^2 - u^2)^{\nu-\mu} \, h(x) \, dx \right\} du.$$

Next, we consider the dual integral equations

$$\int_0^\infty t^{-2\alpha} \, f(t) \, H_\mu(xt) \, dt \; = \; g(x), \; 0 < x < 1, \qquad (2.236)$$

$$\int_0^\infty t^{-2\beta} \, f(t) \, H_\nu(xt) \, dt \; = \; h(x), \; 1 < x < \infty, \qquad (2.237)$$

where $2(\beta - \alpha) = \nu - \mu$, $|\nu| < \frac{1}{2}$ and $|\mu| < \frac{1}{2}$.

To obtain the solution of the above pair, we proceed exactly as above and establish first the solution of the system with $h(x) = 0$, which is

$$f(t) = 2^{\mu-\nu} \, t^{2\beta+1} \int_0^1 u^{\nu-\mu+1} \, Y_\nu(ut) \, I_{\frac{\mu}{2}, \, \nu-\mu}[g] \, du$$

$$\equiv f_1(t), \;\; \text{say},$$

where I is the Erdélyi–Kober operator. Then we set $g(x) = 0$ in (2.236) and find the solution of the system as given by

$$f(t) = 2^{\nu-\mu} \, t^{2\alpha+1} \int_1^\infty u^{\mu-\nu+1} \, Y_\mu(ut) \, I_{\frac{\nu}{2}, \, \mu-\nu}(1, \mu : 2) \, [h] \, du$$

$$\equiv f_2(t), \;\; \text{say},$$

where $I(1, x : 2)$ is the generalized Erdélyi–Kober operator. Therefore,

$$f(t) = f_1(t) + f_2(t)$$

is the solution of the dual integral equations (2.236) and (2.237).

2.3 Dual integral equations related to the Kontorovich−Lebedev transform

In this section, we present some dual integral equations for their solutions which are related to the Kontorovich−Lebedev transforms. These pairs of integral equations arise in the solution of mixed boundary value problems for wedge-shaped regions. Lebedev and Skal'skaya (1974) first studied this class of dual integral equations. They considered the following two pairs of integral equations:

$$\int_0^\infty f(\tau)\,\omega(\tau)\,K_{i\tau}(\lambda r)\,d\tau = rg(r),\ 0 < r < a, \tag{2.238}$$

$$\int_0^\infty f(\tau)\,K_{i\tau}(\lambda r)\,d\tau = h(r),\ a < r < \infty; \tag{2.239}$$

and

$$\int_0^\infty f(\tau)\,K_{i\tau}(\lambda r)\,d\tau = h(r),\ 0 < r < a, \tag{2.240}$$

$$\int_0^\infty f(\tau)\,\omega(\tau)\,K_{i\tau}(\lambda r)\,d\tau = rg(r),\ a < r < \infty. \tag{2.241}$$

Here (r, ϕ, z) is the system of cylindrical coordinates in which the z-axis coincides with the edge of the wedge $(0 < r < \infty,\ -\gamma < \phi < \gamma,\ -\infty < z < \infty)$, $K_{i\tau}(\lambda r)$ is the Macdonald function with imaginary index and $\omega(\tau)$ is the weight function defined by

$$\omega(\tau) = \tau\,\tanh\gamma\tau\ \text{ or }\ \tau\coth\gamma\tau,$$

and λ is a positive real parameter.

2.3.1. Solution of the dual integral equations (2.238), (2.239).

Without any loss of generality, we assume that $g(r) = 0$. We shall seek a solution of this pair in the form

$$f(\tau) = \left(\frac{2}{\pi}\right)^{3/2}\frac{\tau\sinh\pi\tau}{\omega(\tau)}\int_a^\infty \phi(t)\,W^+(\lambda t, i\tau)\,dt, \tag{2.242}$$

with $W^+(x, \nu) = \frac{1}{2}\left\{K_{\frac{1}{2}+\nu}(x) + K_{\frac{1}{2}-\nu}(x)\right\}$, where $\phi(t)$ is an unknown function continuous with its first derivative in $[a, \infty)$ and $\phi(t) \to 0$ as $t \to \infty$. Integrating by parts, (2.242) becomes

$$f(\tau) = -\frac{4}{\pi\sqrt{\pi}} \frac{\tau \sinh \pi\tau}{\omega(\tau)} \phi(a) \int_0^a W^+(\lambda s, i\tau) \, ds$$

$$- \left(\frac{2}{\pi}\right)^{\frac{3}{2}} \frac{\tau \sinh \pi\tau}{\omega(\tau)} \int_a^\infty \phi'(t) \left\{ \int_0^t W^+(\lambda s, \ i\tau) \, ds \right\} dt. \qquad (2.243)$$

Substituting (2.243) into (2.238) with $g(r) = 0$ and then using the result (A.1.31), we see that it is automatically satisfied. Thus, putting (2.242) into equation (2.239), a Fredholm integral equation of the first kind is obtained as

$$\left(\frac{2}{\pi}\right)^{3/2} \int_a^\infty \phi(t) \left\{ \int_0^\infty \frac{\tau \sinh \pi\tau}{\omega(\tau)} W^+(\lambda t, \ i\tau) K_{i\tau}(\lambda r) \, d\tau \right\} dt = h(r), \ a < r < \infty.$$
$$\qquad (2.244)$$

Let us assume that

$$\omega(\tau) = \tau \, \tanh \gamma\tau.$$

Then

$$\frac{\tau \, \sinh \pi\tau}{\omega(\tau)} = \cosh \pi\tau + \frac{\sinh(\pi - \gamma)\tau}{\sinh \gamma\tau}.$$

Using the result (A.1.30), equation (2.244) becomes

$$\int_r^\infty \phi(t) \, \frac{e^{-\lambda(t-r)}}{\sqrt{\lambda(t-r)}} \, dt = H(r), \ a < r < \infty, \qquad (2.245)$$

where

$$H(r) = h(r) - \left(\frac{2}{\pi}\right)^{3/2} \int_a^\infty \phi(s) \left\{ \int_0^\infty \frac{\sinh(\pi - \gamma)\tau}{\sinh \gamma\tau} W^+(\lambda s, i\tau) K_{i\tau}(\lambda r) \, d\tau \right\} ds.$$

By Abel inversion, from (2.245) we get

$$\phi(t) = -\frac{\sqrt{\lambda} \, e^{\lambda t}}{\pi} \frac{d}{dt} \int_t^\infty \frac{e^{-\lambda r}}{\sqrt{r-t}} H(r) \, dr. \qquad (2.246)$$

Utilizing the relation

$$W^+(\lambda t, \ i\tau) = -\frac{e^{\lambda t}}{\sqrt{2\pi\lambda}} \frac{d}{dt} \int_t^\infty \frac{e^{-\lambda r} K_{i\tau}(\lambda r)}{\sqrt{r-t}} \, dr \qquad (2.247)$$

which can be obtained by inverting (A.1.31) and then differentiating with respect to t, equation (2.246) reduces to a Fredholm integral equation of the second kind with symmetric kernel as given by

$$\phi(t) = -\frac{\sqrt{\lambda}\, e^{\lambda t}}{\pi} \frac{d}{dt} \int_t^\infty \frac{e^{-\lambda r}\, h(r)}{\sqrt{r-t}}\, dr$$

$$-\frac{\lambda}{\pi} \int_a^\infty \phi(s)\, K(s,t)\, ds,\ a \le t < \infty, \qquad (2.248)$$

where

$$K(s,t) = \frac{4}{\pi} \int_0^\infty \frac{\sinh(\pi-\gamma)\tau}{\sinh\gamma\tau}\, W^+(\lambda s,\ i\tau)\, W^+(\lambda t,\ i\tau)\, d\tau,\ 0 < \gamma \le \pi.$$

Similarly, assuming $\omega(\tau) = \tau \coth\gamma\tau$, we obtain the Fredholm integral equation of the second kind as

$$\phi(t) = -\frac{\sqrt{\lambda}\, e^{\lambda t}}{\pi} \frac{d}{dt} \int_t^\infty \frac{e^{-\lambda r} h(r)}{\sqrt{r-t}}$$

$$+\frac{\lambda}{\pi} \int_a^\infty \phi(s)\, K(s,t)\, dt,\ a \le t < \infty, \qquad (2.249)$$

where

$$K(s,t) = \frac{4}{\pi} \int_0^\infty \frac{\cosh(\pi-\gamma)\tau}{\cosh\gamma\tau}\, W^+(\lambda s,\ i\tau)\, W^+(\lambda t,\ i\tau)\, d\tau,\ 0 < \gamma \le \pi.$$

The integral equations (2.248) and (2.249) can be solved by an iteration process which converges rapidly when values of λa are not too small. In particular, when $\frac{\pi}{2} \le \gamma \le \pi$, we have

$$K^2(s,t) \le \{K_0(2\lambda s) + K_1(2\lambda s)\} \{K_0(2\lambda t) + K_1(2\lambda t)\},$$

$$\|K(s,t)\| \le \frac{1}{\lambda} K_0(2\lambda a).$$

From the general theory of integral equations, we find that the condition for convergence of the iteration process is

$$K_0(2\lambda a) < \pi.$$

The above inequality holds when $\lambda a > 0.025$ and the rate of convergence of the iteration process increases with increasing λa. After obtaining the solutions of these integral equations, the solution of the dual integral equations is obtained by using relation (2.242).

2.3.2 Solution of the dual integral equations (2.240), (2.241).

To find the solution of this pair, we assume that $h(r) = 0$. Let us assume that the solution of this pair can be expressed as

$$f(\tau) = \left(\frac{2}{\pi}\right)^{\frac{3}{2}} \sinh \pi\tau \int_a^\infty \phi(t) \; W^-(\lambda t, \; i\tau) \; dt, \tag{2.250}$$

with $W^-(x, \nu) = \frac{1}{2}\left\{K_{\frac{1}{2}+\nu}(x) - K_{\frac{1}{2}-\nu}(x)\right\}$, where $\phi(t)$ is an unknown function which is continuous together with its first derivative in $[a, \infty)$ and $\phi(t) \to 0$ as $t \to \infty$. Using the result (A.1.32), we observe that equation (2.240) (with $h(r) = 0$) is satisfied identically.

Equation (2.250) can be written as

$$f(\tau) = \left(\frac{2}{\pi}\right)^{3/2} \sinh \pi\tau \int_a^\infty \phi(t) \; d\left\{\int_0^t W^+(\lambda s, \; i\tau) \; ds\right\}.$$

Then integrating by parts and substituting it into (2.241), we get

$$-\left(\frac{2}{\pi}\right)^{\frac{3}{2}} \phi(a) \int_0^\infty \sinh \pi\tau \; \omega(\tau) \left\{\int_0^a W^-(\lambda s, \; i\tau) \; ds\right\} K_{i\tau}(\lambda r) \; d\tau$$

$$-\left(\frac{2}{\pi}\right)^{\frac{3}{2}} \int_a^\infty \phi'(t) \left[\int_0^\infty \sinh \pi\tau \; \omega(\tau) \left\{\int_0^t W^-(\lambda s, \; i\tau) \; ds\right\} K_{i\tau}(\lambda r) \; d\tau\right] dt$$

$$= r g(r), \; a < r < \infty. \tag{2.251}$$

When $\omega(\tau) = \tau \tanh \gamma\tau$, then

$$\sinh \pi\tau \; \omega(\tau) = \tau \; \cosh \pi\tau - \tau \; \frac{\cosh(\pi - \gamma)\tau}{\cosh \gamma\tau},$$

and thus using the result (A.1.33), equation (2.251) can be reduced to

$$-\frac{d}{dr}\left\{e^{-\lambda r} \int_r^\infty \frac{e^{-\lambda(t-r)}}{\sqrt{\lambda(t-r)}} \phi(t) \; dt\right\} = e^{-\lambda r} \; g(r) + \left(\frac{2}{\pi}\right)^{\frac{3}{2}} e^{-\lambda r}$$

$$\times \int_a^\infty \phi(s) \left\{\int_0^\infty \tau \; \frac{\cosh(\pi - \gamma)\tau}{\cosh \gamma\tau} W^-(\lambda s, \; i\tau) \frac{K_{i\tau}(\lambda r)}{r} \; d\tau\right\} ds, \; a < r < \infty. \tag{2.252}$$

64

Following the relation

$$\frac{\tau\,e^{\lambda t}}{\sqrt{2\pi\lambda}}\int_t^\infty \frac{e^{-\lambda r}K_{i\tau}(\lambda r)}{r\sqrt{r-t}}\,dr = W^-(\lambda t,\; i\tau)$$

which can be derived from (A.1.32), equation (2.252) reduces to the following Fredholm integral equation of the second kind

$$\phi(t) = \frac{\sqrt{\lambda}\,e^{\lambda t}}{\pi}\int_t^\infty \frac{e^{-\lambda r}g(r)}{\sqrt{r-t}}\,dr$$
$$+\frac{\lambda}{\pi}\int_a^\infty \phi(s)\,K(s,t)\,ds,\; a\le t<\infty, \qquad (2.253)$$

where

$$K(s,t) = \frac{4}{\pi}\int_0^\infty \frac{\cosh(\pi-\gamma)\tau}{\cosh\gamma\tau}\,W^-(\lambda s, i\tau)\,W^-(\lambda t, i\tau)\,d\tau,\; 0<\gamma\le\pi. \qquad (2.254)$$

In a manner similar to the above, when $\omega(\tau) = \tau\coth\gamma\tau$, the Fredholm integral equation of the second kind is

$$\phi(t) = \frac{\sqrt{\lambda}\,e^{\lambda t}}{\pi}\int_t^\infty \frac{e^{-\lambda r}g(r)}{\sqrt{r-t}}\,dr - \frac{\lambda}{\pi}\int_a^\infty \phi(s)\,K(s,t)\,ds,\; a\le t<\infty, \qquad (2.255)$$

where

$$K(s,t) = \frac{4}{\pi}\int_0^\infty \frac{\sinh(\pi-\gamma)\tau}{\sinh\gamma\tau}\,W^-(\lambda s,\; i\tau)\,W^-(\lambda t,\; i\tau)\,d\tau,\; 0<\gamma\le\pi. \qquad (2.256)$$

An Application

As an application, we consider the problem of constructing a harmonic function $u = u(r,\phi,z)$ in the region $0<r<\infty$, $-\gamma<\phi<\gamma$, $0<z<z_0$ with the mixed boundary conditions

$$u\big|_{z=0} = u\big|_{z=z_0} = 0,\; \frac{1}{r}\frac{\partial u}{\partial\phi}\Big|_{\substack{\phi=\pm\gamma\\ r<a}} = 0,\; u\big|_{\substack{\phi=\pm\gamma\\ r>a}} = h(r,z).$$

The solution of the above mixed boundary value problem is given by the representation

$$u = \sum_{n=1}^\infty \sin\frac{n\pi z}{z_0}\int_0^\infty f_n(\tau)\,\frac{\cosh\phi\tau}{\cosh\gamma\tau}\,K_{i\tau}\left(\frac{n\pi r}{z_0}\right)d\tau$$

where $f_n(\tau)$ satisfies the dual integral equations (2.238) and (2.239) with $\omega(\tau) = \tau\tanh\gamma\tau$, $\lambda = \frac{n\pi}{z_0}$, $h(r) = h_n(r)$, where $h_n(r)$ are the coefficients of the Fourier expansion of the function $h(r,z)$, and $g(r) = 0$.

2.4 Dual integral equations associated with inverse Weber−Orr transforms

This section is concerned with some dual integral equations involving inverse Weber −Orr transforms. Srivastava (1964a) first encountered this type of dual integral equation in the study of axisymmetric deformation of an infinite elastic solid containing an exterior plane crack with a circular boundary and an infinitely long cylindrical cavity, the axis of the cylinder being normal to the plane of the crack and passing through the centre of the circle bounding the crack.

The problem is to determine a harmonic function $\phi(r, z)$ which satisfies Laplace equation

$$\frac{\partial^2 \phi}{\partial r^2} + \frac{1}{r} \frac{\partial \phi}{\partial r} + \frac{\partial^2 \phi}{\partial z^2} = 0, \tag{2.257}$$

in the region $1 < r < \infty$, $0 < z < \infty$ such that $\phi(r, z) \to 0$ as $\sqrt{r^2 + z^2} \to \infty$ with the following boundary conditions:

$$\phi(1, z) = 0, \; 0 \leq z < \infty, \tag{2.258}$$

$$\phi(r, 0) = 0, \; 1 \leq r \leq a, \tag{2.259}$$

$$\frac{\partial \phi}{\partial z} = -g(r), \; z = 0, \; a < r < \infty. \tag{2.260}$$

The above mixed boundary value problem can be reduced to the solution of a pair of integral equations by applying the Weber integral transform. We use here the Weber transform $\overline{\phi}(s)$ of a continuous real function $\phi(r)$ defined on $(1, \infty)$ together with its inversion given by (see Appendix A.1.4)

$$W[\phi] \equiv \overline{\phi}(s) = \int_1^\infty r \left\{ J_0(rs) Y_0(s) - Y_0(rs) J_0(s) \right\} \phi(r) \, dr, \tag{2.261}$$

$$W^{-1}[\overline{\phi}] \equiv \phi(r) = \int_0^\infty \frac{\left\{ J_0(rs) Y_0(s) - Y_0(rs) J_0(s) \right\}}{J_0^2(s) + Y_0^2(s)} \overline{\phi}(s) \, ds. \tag{2.262}$$

Taking the Weber transform of equation (2.257), we get

$$\frac{d^2 \overline{\phi}}{dz^2} - s^2 \overline{\phi} = 0 \tag{2.263}$$

whose solution is

$$\overline{\phi}(s, z) = f(s) \, e^{-sz} + \psi(s) \, e^{sz}, \tag{2.264}$$

where $f(s)$, $\psi(s)$ are unknown functions to be determined. Using the inversion formula (2.262), relation (2.264) gives

$$\phi(r, z) = \int_0^\infty s \, \frac{\{J_0(rs)Y_0(s) - Y_0(rs)J_0(s)\}}{J_0^2(s) + Y_0^2(s)} \, [f(s) \, e^{-sz} + \psi(s) \, e^{sz}] \, ds. \tag{2.265}$$

Since $\phi(r, z) \to 0$ as $\sqrt{r^2 + z^2} \to \infty$, we must have $\psi(s) = 0$. Condition (2.258) is automatically satisfied for all $f(s)$. To satisfy the boundary conditions (2.259) and (2.260), $f(s)$ satisfies the dual integral equations

$$W^{-1}[f(s)] = \int_0^\infty s \frac{\{J_0(rs)Y_0(s) - Y_0(rs)J_0(s)\}}{J_0^2(s) + Y_0^2(s)} \, f(s) \, ds = 0, \; 1 < r < a, \tag{2.266}$$

$$W^{-1}[sf(s)] = \int_0^\infty s^2 \frac{\{J_0(rs)Y_0(s) - Y_0(rs)J_0(s)\}}{J_0^2(s) + Y_0^2(s)} \, f(s) \, ds = g(r), \; a < r < \infty. \tag{2.267}$$

2.4.1 Solution of the pair (2.266) and (2.267).

To solve the above pair of integral equations, we put

$$f(s) = W \, [F(r)] \,, \tag{2.268}$$

where

$$F(r) \;=\; 0, \; 1 \le r \le a, \tag{2.269}$$

$$F(r) \;=\; \int_a^r \frac{\psi(t) \, dt}{\sqrt{r^2 - t^2}}, \; a < r < \infty, \tag{2.270}$$

and $\psi(t)$ is an unknown bounded and continuously differentiable real auxiliary function defined in $[a, \infty)$. Then equation (2.268) can be expressed as

$$f(s) = \int_a^\infty r \, \{J_0(rs) \, Y_0(s) - Y_0(rs) \, J_0(s)\} \left(\int_a^r \frac{\psi(t) \, dt}{\sqrt{r^2 - t^2}} \right) dr. \tag{2.271}$$

Interchanging the order of integration in the above relation and then using two known relations (cf. Erdélyi et al. 1954b, (7), p. 7 and (33), p. 103), we obtain

$$sf(s) = Y_0(s) \int_a^\infty \psi(t) \cos st \, dt - J_0(s) \int_a^\infty \phi(t) \, \sin st \, dt. \tag{2.272}$$

67

Integrating by parts, (2.272) becomes

$$s^2 f(s) = -\psi(a) \left[J_0(s) \, \cos as + Y_0(s) \, \sin as \right]$$

$$-\int_a^\infty \psi'(t) \left[Y_0(s) \, \sin st + J_0(s) \, \cos st \right] dt$$

$$= -\frac{1}{2} \psi(a) \left[H_0^{(1)}(s) \, e^{-ias} + H_0^{(2)}(s) \, e^{ias} \right]$$

$$-\frac{1}{2} \int_a^\infty \psi'(t) \left[H_0^{(1)}(s) \, e^{-its} + H_0^{(2)}(s) \, e^{its} \right] dt, \qquad (2.273)$$

where $H_0^{(1)}(s)$ and $H_0^{(2)}(s)$ are the Bessel functions of the third kind defined by the relations

$$H_0^{(1)}(s) = J_0(s) + i \, Y_0(s),$$
$$H_0^{(2)}(s) = J_0(s) - i \, Y_0(s).$$

Now

$$\frac{J_0(rs) \, Y_0(s) - Y_0(rs) \, J_0(s)}{J_0^2(s) + Y_0^2(s)} = -\frac{1}{2i} \left[\frac{H_0^{(1)}(rs)}{H_0^{(1)}(s)} - \frac{H_0^{(2)}(rs)}{H_0^{(2)}(s)} \right].$$

Therefore, using equation (2.273) the left side of (2.267) becomes

$$\int_0^\infty s^2 f(s) \, \frac{\{ J_0(rs) \, Y_0(s) - Y_0(rs) \, J_0(s) \}}{J_0^2(s) + Y_0^2(s)} \, ds$$

$$= \frac{1}{4i} \left[\psi(a) \int_0^\infty \left\{ H_0^{(1)}(rs) \, e^{-ias} - H_0^{(2)}(rs) \, e^{ias} \right\} ds \right.$$

$$+ \psi(a) \int_0^\infty \left\{ \frac{H_0^{(1)}(rs) H_0^{(2)}(s)}{H_0^{(1)}(s)} e^{ias} - \frac{H_0^{(1)}(s) H_0^{(2)}(rs)}{H_0^{(2)}(s)} e^{-ias} \right\} ds$$

$$+ \int_a^\infty \psi'(t) \left\{ \int_t^\infty \left(H_0^{(1)}(rs) \, e^{-ist} - H_0^{(2)}(rs) \, e^{ist} \right) ds \right\} dt$$

$$+ \int_a^\infty \psi'(t) \left\{ \int_t^\infty \left(\frac{H_0^{(1)}(rs) H_0^{(2)}(s)}{H_0^{(1)}(s)} e^{ist} - \frac{H_0^{(2)}(rs) H_0^{(1)}(s)}{H_0^{(2)}(s)} e^{-ist} \right) ds \right\} dt \right]$$

$$(2.274)$$

Since for real s, $H_0^{(2)}(s)$ is the conjugate of $H_0^{(1)}(s)$, (2.274) can be written as

$$\int_0^\infty s^2 f(s) \, \frac{\{J_0(rs) \, Y_0(s) - Y_0(rs) \, J_0(s)\}}{J_0^2(s) + Y_0^2(s)} \, ds$$

$$= \frac{1}{2} \psi(a) \int_0^\infty \{Y_0(rs) \, \cos as - J_0(rs) \, \sin as\} \, ds$$

$$+ \frac{1}{2} \psi(a) \, Im \left(\int_0^\infty \frac{H_0^{(1)}(rs) \, H_0^{(2)}(s)}{H_0^{(1)}(s)} \, e^{ias} \, ds \right)$$

$$+ \frac{1}{2} \int_a^\infty \psi'(t) \left\{ \int_0^\infty (Y_0(rs) \, \cos st - J_0(rs) \, \sin st) \, ds \right\} dt$$

$$+ \frac{1}{2} \int_a^\infty \psi'(t) \, Im \left(\int_0^\infty \frac{H_0^{(1)}(rs) \, H_0^{(2)}(s)}{H_0^{(1)}(s)} \, e^{ist} \, ds \right) dt. \tag{2.275}$$

Now considering the contour integral

$$\int_\Gamma \frac{H_0^{(1)}(rz)}{H_0^{(1)}(z)} \, J_0(z) \, e^{izt} \, dz,$$

where Γ is the contour (suitably indented at the origin) consisting of the positive real axis, the arc of the first quadrant and the positive imaginary axis, it is easily shown that

$$Im \left(\int_0^\infty \frac{H_0^{(1)}(rs)}{H_0^{(1)}(s)} J_0(s) \, e^{ist} \, ds \right) = \int_0^\infty I_0(\xi) \frac{K_0(r\xi)}{K_0(\xi)} \, e^{-\xi t} \, d\xi,$$

where $I_0(\xi)$ is the modified Bessel function of the first kind. Then

$$Im \left(\int_0^\infty \frac{H_0^{(1)}(rs) H_0^{(2)}(s)}{H_0^{(1)}(s)} \, e^{ist} \, ds \right) = Im \left(\int_0^\infty \frac{\{2J_0(s) - H_0^{(1)}(s)\}}{H_0^{(1)}(s) \, e^{-ist}} H_0^{(1)}(rs) \, ds \right)$$

$$= 2 \int_0^\infty I_0(\xi) \frac{K_0(r\xi)}{K_0(\xi)} \, e^{-\xi t} \, d\xi - \int_0^\infty \{J_0(rs) \, \sin st + Y_0(rs) \, \cos st\} \, ds. \tag{2.276}$$

Using the relations (cf. Erdélyi et al. 1954a, (28), p. 47 and (1), p. 99)

$$\int_0^\infty J_0(ax) \, \sin xy \, dx = - \int_0^\infty Y_0(ax) \, \cos xy \, dx$$

$$= \begin{cases} 0, & 0 < y < a, \\ (y^2 - a^2)^{-\frac{1}{2}}, & a < y < \infty, \end{cases}$$

69

and equation (2.276), (2.275) reduces to

$$\int_0^\infty s^2 f(s) \, \frac{\{J_0(rs) \, Y_0(s) - Y_0(rs) \, J_0(s)\}}{J_0^2(s) + Y_0^2(s)} \, ds$$

$$= -\frac{\psi(a)}{\sqrt{a^2 - r^2}} \, H(a - r) + \psi(a) \int_0^\infty I_0(\xi) \, \frac{K_0(\xi r)}{K_0(\xi)} \, e^{-a\xi} \, d\xi$$

$$- \int_r^\infty \frac{\psi'(t) \, dt}{\sqrt{t^2 - r^2}} + \int_a^\infty \psi'(t) \left\{ \int_0^\infty I_0(\xi) \, \frac{K_0(\xi r)}{K_0(\xi)} \, e^{-\xi t} \, d\xi \right\} dt, \qquad (2.277)$$

where $H(x)$ is the Heaviside unit function.

Therefore, from equations (2.267) and (2.277) we get an Abel integral equation of the second kind for the unknown function $\psi(t)$ as given by

$$\int_r^\infty \frac{\psi'(t) \, dt}{\sqrt{t^2 - r^2}} = G(r), \ a < r < \infty, \qquad (2.278)$$

where

$$G(r) = -g(r) + \int_a^\infty \psi(u) \left\{ \int_0^\infty \frac{\xi \, I_0(\xi) \, K_0(\xi r)}{K_0(\xi)} \, e^{-\xi u} \, d\xi \right\} du. \qquad (2.279)$$

The solution of the integral equation (2.278) is

$$\psi'(t) = -\frac{2}{\pi} \frac{d}{dt} \int_t^\infty \frac{r \, G(r)}{\sqrt{r^2 - t^2}} \, dr.$$

Since $\psi(t) \to 0$ as $t \to \infty$, we get

$$\psi(t) = -\frac{2}{\pi} \int_t^\infty \frac{r \, G(r)}{\sqrt{r^2 - t^2}} \, dr. \qquad (2.280)$$

Using the relation (cf. Erdélyi et al. 1954b, (13), p. 129)

$$\int_a^\infty \frac{x \, K_0(xy)}{\sqrt{x^2 - a^2}} \, dx = \frac{\pi}{2y} \, e^{-ay}, \qquad (2.281)$$

equation (2.280) reduces to a Fredholm integral equation of the second kind as given by

$$\psi(t) + \int_a^\infty \psi(u) \left\{ \int_0^\infty \frac{I_0(\xi)}{K_0(\xi)} \, e^{-\xi(u+t)} \, d\xi \right\} du$$

$$= \frac{2}{\pi} \int_t^\infty \frac{r \, g(r)}{\sqrt{r^2 - t^2}} \, dr, \ a \le t < \infty. \qquad (2.282)$$

Therefore, the solution of the dual integral equations (2.266), and (2.267) is given by (2.271) after obtaining the solution of the above Fredholm integral equation (2.282) for the unknown function $\psi(t)$.

2.4.2 Srivastav (1964b) considered the more general dual integral equations (also see Virchenko (1989), p. 62)

$$\int_0^\infty t\, f(t)\frac{\{J_\nu(xt)\, Y_\nu(t) - Y_\nu(xt)\, J_\nu(t)\}}{J_\nu^2(t) + Y_\nu^2(t)}\, dt \;= f_1(x),\; 1 < x < a,\quad (2.283)$$

$$\int_0^\infty t^{1+2\alpha}\, f(t)\frac{\{J_\nu(xt)\, Y_\nu(t) - Y_\nu(xt)\, J_\nu(t)\}}{J_\nu^2(t) + Y_\nu^2(t)}\, dt \;= -f_2(x),\; a < x < \infty,\quad (2.284)$$

where $-\frac{1}{2} \le \alpha \le \frac{1}{2}$ with $\alpha \ne 0$. The solution of this pair is obtained for two cases $f_1(x) = 0$ and $f_2(x) = 0$. The solution to the general problem can obviously be obtained by adding the two solutions. Before obtaining the solution of this pair, we present some integral properties of Bessel functions which are needed here.

Defining

$$R_{\mu,\nu}(x,t) \equiv J_\mu(xt)\, Y_\nu(t) - Y_\mu(xt)\, J_\nu(t),\quad (2.285)$$

we can express it as

$$R_{\mu,\nu}(x,t) = -\frac{1}{2i}\left(\frac{H_\mu^{(1)}(xt)}{H_\nu^{(1)}(t)} - \frac{H_\mu^{(2)}(xt)}{H_\nu^{(2)}(t)}\right)\, H_\nu^{(1)}(t)\, H_\nu^{(2)}(t).\quad (2.286)$$

From the formulae (32), p. 25 and (38), p. 104 of Erdélyi et al. (1954b) it is found that

$$\int_u^\infty x^{1-\mu}\, R_{\mu,\nu}(x,t)\, (x^2 - u^2)^\alpha\, dx = 2^\alpha\, \Gamma(\alpha+1)\, u^{1+\alpha-\mu}\, t^{-\alpha-1}\, R_{\mu-\alpha-1,\nu}(u,t),\quad (2.287)$$

where $Re(\alpha) > -1$ and $Re(\mu - 2\alpha) > \frac{1}{2}$.

The Weber−Orr transform and its inversion are given by (see Appendix A.1.4)

$$W_{\nu,\nu}[g] \;\equiv\; \overline{g}(t) = \int_1^\infty x\, g(x)\, R_{\nu,\nu}(x,t)\, dx,\; 0 < t < \infty,\quad (2.288)$$

$$W_{\nu,\nu}^{-1}[\overline{g}] \;\equiv\; g(x) = \int_0^\infty \frac{t\, \overline{g}(t)\, R_{\nu,\nu}(x,t)}{H_\nu^{(1)}(t)\, H_\nu^{(2)}(t)}\, dt,\; 1 < x < \infty.\quad (2.289)$$

From relations (2.287) and (2.289), we obtain that

$$\int_0^\infty t^{-\alpha} \frac{R_{\nu-\alpha-1,\nu}(u,t)\ R_{\nu,\nu}(x,t)}{H_\nu^{(1)}(t)\ H_\nu^{(2)}(t)}\ dt = \begin{cases} 0, & 1 < x < u, \\ \dfrac{2^{-\alpha}u^{\nu-\alpha-1}(x^2-u^2)^\alpha}{\Gamma(\alpha+1)\ x^\nu}, & x > u. \end{cases} \quad (2.290)$$

It can easily be shown that

$$\int_0^\infty t^{\alpha+1} \frac{R_{\eta,\nu}(x,t)\ R_{\mu,\nu}(u,t)}{H_\nu^{(1)}(t)\ H_\nu^{(2)}(t)}\ dt$$

$$= Re\left(\int_0^\infty t^{\alpha+1} \frac{H_\eta^{(1)}(xt)\ H_\mu^{(1)}(ut)}{H_\nu^{(1)}(t)}\ J_\nu(t)\ dt\right) + \int_0^\infty t^{\alpha+1}\ J_\eta(xt)\ J_\mu(ut)\ dt. \quad (2.291)$$

Considering the contour integral

$$\int_\Gamma z^{\alpha+1}\ J_\nu(z)\ \frac{H_\eta^{(1)}(xz)\ H_\mu^{(1)}(uz)}{H_\nu^{(1)}(z)}dz,$$

where Γ consists of the positive real axis, the arc of the first quadrant and the positive imaginary axis, we find that

$$Re\left(\int_0^\infty t^{\alpha+1}\ \frac{J_\nu(t)\ H_\eta^{(1)}(xt)\ H_\mu^{(1)}(ut)}{H_\nu^{(1)}(t)}\ dt\right)$$

$$= \frac{2}{\pi}\cos\left[(2\nu+\alpha+1-\eta-\mu)\frac{\pi}{2}\right]\int_0^\infty s^{\alpha+1}\ \frac{I_\nu(s)}{K_\nu(s)}\ K_\eta(sx)\ K_\mu(us)\ ds. \quad (2.292)$$

First we consider the pair

$$\int_0^\infty t\ f(t)\ \frac{\{J_\nu(xt)Y_\nu(t) - Y_\nu(xt)J_\nu(t)\}}{J_\nu^2(t)+Y_\nu^2(t)}\ dt = 0,\ 1 < x < a, \quad (2.293)$$

$$\int_0^\infty t^{1+2\alpha}\ f(t)\ \frac{\{J_\nu(xt)Y_\nu(t) - Y_\nu(xt)J_\nu(t)\}}{J_\nu^2(t)+Y_\nu^2(t)}\ dt = -f_2(x),\ a < x < \infty. (2.294)$$

The solution of this pair is obtained in two cases.

Case I: $-\frac{1}{2} \leq \alpha < 0$.

We assume that

$$W_{\nu,\nu}^{-1}[f] = x^{-\nu-1}\ \frac{\partial}{\partial x}\int_a^x g(u)\ (x^2-u^2)^\alpha\ du,\ x > a, \quad (2.295)$$

where $g(u)$ is an unknown function. From the relations (2.287) and (2.295), we get

$$f(t) = 2^\alpha \; \Gamma(\alpha + 1) \; t^{-\alpha} \int_a^\infty g(u) \; u^{\alpha - \nu} \; R_{\nu - \alpha, \nu}(u, t) \; du. \qquad (2.296)$$

The above expression for $f(t)$ satisfies (2.293) identically. Then from (2.294) and (2.296), interchanging the order of integration, we have

$$2^\alpha \; \Gamma(\alpha + 1) \int_a^\infty g(u) \; u^{\alpha - \nu} \; H_{\nu, \nu, \nu - \alpha, \alpha}(x, u) \; du = -f_2(x), \; x > a, \qquad (2.297)$$

where

$$H_{\nu, \eta, \mu, \alpha}(x, t) \equiv \int_0^\infty \xi^{\alpha + 1} \frac{R_{\eta, \nu}(x, \xi) \; R_{\mu, \nu}(t, \xi)}{H_\nu^{(1)}(\xi) \; H_\nu^{(2)}(\xi)} \; d\xi. \qquad (2.298)$$

Using the relations (A.1.18), (2.291) and (2.292), we obtain

$$H_{\nu, \nu, \nu - \alpha, \alpha}(x, u) \;=\; \frac{2^{\alpha + 1} \; x^\nu \; (u^2 - x^2)^{-\alpha - 1}}{\Gamma(-\alpha) \; u^{\nu - \alpha}} \; H(u - x)$$

$$-\frac{2}{\pi} \sin \pi \alpha \int_0^\infty s^{\alpha + 1} \frac{I_\nu(s)}{K_\nu(s)} \; K_{\nu - \alpha}(us) \; K_\nu(xs) \; ds, \quad (2.299)$$

where $H(x)$ is the Heaviside unit function. Substituting the above expression into (2.297), then by Abel inversion together with the use of the relation for $Re(\mu) > -1$, $Re(s) > 0$,

$$\int_a^\infty x^{1-\nu} (x^2 - a^2)^\mu \; K_\nu(xs) \; dx = 2^\mu \; a^{\mu - \nu + 1} \; s^{-\mu - 1} \; \Gamma(1 + \mu) \; K_{\mu - \nu + 1}(as), \qquad (2.300)$$

a Fredholm integral equation of the second kind is obtained as

$$g(u) \; u^{\alpha - \nu} + \frac{2u}{\pi} \sin \pi \alpha \int_a^\infty g(v) \; v^{\alpha - \nu} \left\{ \int_0^\infty \frac{s \; I_\nu(s)}{K_\nu(s)} \; K_{\nu - \alpha}(vs) \; K_{\alpha - \nu}(us) \; ds \right\} dv$$

$$= \frac{2^{-2\alpha} \; u^{\nu - \alpha}}{\{\Gamma(1 + \alpha)\}^2} \frac{d}{du} \int_u^\infty x^{1-\nu} (x^2 - u^2)^\alpha \; f_2(x) \; dx. \qquad (2.301)$$

Case II: $0 < \alpha \leq \frac{1}{2}$ and $\nu > 0$.

The case for $\nu = 0$ has already been discussed in §2.4.1. For $\nu > 0$ and $0 < \alpha \leq \frac{1}{2}$, we set

$$W_{\nu, \nu}^{-1}[f] = x^{-\nu} \int_a^x (x^2 - u^2)^{\alpha - 1} \; g(u) \; du, \; x > a. \qquad (2.302)$$

Then, we have

$$f(t) = 2^{\alpha-1}\,\Gamma(\alpha)\,t^{-\alpha}\int_a^\infty u^{\alpha-\nu}\,g(u)\,R_{\nu-\alpha,\nu}(u,t)\,du. \qquad (2.303)$$

Multiplying both sides of equation (2.294) by $x^{1-\nu}$ and integrating from x to ∞, then using (2.303) together with the interchange of order of integration, we get

$$2^{\alpha-1}\,\Gamma(\alpha)\int_a^\infty u^{\alpha-\nu}\,g(u)\,H_{\nu,\nu-1,\nu-\alpha,\alpha-1}(x,u)\,du$$

$$= -x^{\nu-1}\int_x^\infty s^{1-\nu}f_2(s)\,ds,\ x > a. \qquad (2.304)$$

From the results (A.1.18), (2.291) and (2.292), it can be shown that

$$H_{\nu,\nu-1,\nu-\alpha,\alpha-1}(x,u) = \frac{2^\alpha x^{\nu-1} u^{\alpha-\nu}(u^2-x^2)^{-\alpha}}{\Gamma(1-\alpha)}\,H(u-x)$$

$$-\frac{2}{\pi}\sin\pi\alpha\int_0^\infty s^\alpha\,\frac{K_{\nu-1}(xs)\,K_{\nu-\alpha}(su)\,I_\nu(s)}{K_\nu(s)}\,ds. (2.305)$$

Using the above expression, after some simplification (2.304) becomes

$$g(u)u^{\alpha-\nu} + \frac{2u}{\pi}\sin\ \pi\alpha\int_a^\infty v^{\alpha-\nu}\,g(v)\left\{\int_0^\infty s\,\frac{I_\nu(s)}{K_\nu(s)}\,K_{\nu-\alpha}(vs)\,K_{\alpha-\nu}(us)\,ds\right\}dv$$

$$= \frac{2^{1-2\alpha}\,u^{\nu-\alpha}}{\{\Gamma(1-\alpha)\}^2}\,\frac{d}{du}\int_u^\infty s^{1-\nu}\,(s^2-u^2)^\alpha\,f_2(s)\,ds. \qquad (2.306)$$

This is a Fredholm integral equation of the second kind.

Next, we consider the pair of integral equations

$$\int_0^\infty t\,f(t)\,\frac{\{J_\nu(xt)\,Y_\nu(t) - Y_\nu(xt)\,J_\nu(t)\}}{J_\nu^2(t) + Y_\nu^2(t)}\,dt = f_1(x),\ 1 < x < a, \quad (2.307)$$

$$\int_0^\infty t^{1+2\alpha}\,f(t)\,\frac{\{J_\nu(xt)\,Y_\nu(t) - Y_\nu(xt)\,J_\nu(t)\}}{J_\nu^2(t) + Y_\nu^2(t)}\,dt = 0,\ a < x < \infty. \quad (2.308)$$

Case I: $-\frac{1}{2} \le \alpha < 0$.

We assume that

$$W_{\nu,\nu}^{-1}[f] = x^{-\nu-1}\frac{\partial}{\partial x}\int_a^x(u^2-x^2)^\alpha\,g(u)\,du,\ x > a. \qquad (2.309)$$

Then

$$f(t) = 2^\alpha \, \Gamma(\alpha+1) \, t^{-\alpha} \int_a^\infty u^{\alpha-\nu} \, g(u) \, R_{\nu-\alpha,\nu}(u,t) \, du$$

$$+ \int_1^a x \, f_1(x) \, R_{\nu,\nu}(x,t) \, dx. \tag{2.310}$$

Substituting the above expression for $f(t)$ into (2.308) and then interchanging the order of integration, we find that

$$2^\alpha \, \Gamma^{\alpha+1} \int_a^\infty u^{\alpha-\nu} \, g(u) \, H_{\nu,\nu,\nu-\alpha,\alpha}(x,t) \, dt$$

$$= - \int_1^a s \, f_1(s) \, H_{\nu,\nu,\nu,2\alpha-1}(s,x) \, ds, \ x > a, \tag{2.311}$$

which is of the same form as (2.297). Therefore, it can be reduced to the Fredholm integral equation of the second kind

$$g(u) \, u^{\alpha-\nu} + \frac{2u}{\pi} \, \sin \pi\alpha \int_a^\infty v^{\alpha-\nu} g(v) \left\{ \int_0^\infty s \, \frac{I_\nu(s)}{K_\nu(s)} \, K_{\nu-\alpha}(vs) \, K_{\nu-\alpha}(us) \, ds \right\} \, dv$$

$$= -\frac{2^{-\alpha} \, u}{\Gamma(1+\alpha)} \int_1^a s \, f_1(s) \, H_{\nu,\nu,\nu-\alpha,\alpha-1}(s,u) \, ds, \ a < u < \infty. \tag{2.312}$$

The case $0 < \alpha \le \frac{1}{2}$ can also be treated in a similar manner and results in the same form of the integral equation presented in the previous case.

Now we consider a more general pair of integral equations

$$\int_0^\infty \frac{t \, f(t) \, R_{\nu,\nu}(x,t)}{J_\nu^2(t) + Y_\nu^2(t)} \, dt = 0, \ 1 < x < a, \tag{2.313}$$

$$\int_0^\infty t^{1+2\alpha} \, \{1 + \omega(t)\} \, f(t) \, \frac{R_{\nu,\nu}(x,t)}{J_\nu^2(t) + Y_\nu^2(t)} \, dt = -f_2(x), \ x > a, \tag{2.314}$$

where $-\frac{1}{2} \le \alpha \le \frac{1}{2}$ ($\alpha \ne 0$), $\omega(t)$ is a known weight function.

The equation (2.314) can be expressed as

$$\int_0^\infty t^{1+2\alpha} \frac{f(t) \, R_{\nu,\nu}(x,t)}{J_\nu^2(t) + Y_\nu^2(t)} \, dt = -F_2(x), \ x > a, \tag{2.315}$$

where

$$F_2(x) = f_2(x) + \int_0^\infty t^{1+2\alpha} \, \omega(t) \, f(t) \, \frac{R_{\nu,\nu}(x,t)}{J_\nu^2(t) + Y_\nu^2(t)} \, dt. \tag{2.316}$$

For $-\frac{1}{2} \leq \alpha < 0$,

$$f(t) = 2^\alpha \, \Gamma(\alpha + 1) \, t^{-\alpha} \int_a^\infty u^{\alpha-\nu} \, g(u) \, R_{\nu-\alpha,\nu}(u,t) \, du, \qquad (2.317)$$

and for $0 < \alpha \leq \frac{1}{2}$,

$$f(t) = 2^{\alpha-1} \, \Gamma(\alpha) \, t^{-\alpha} \int_a^\infty u^{\alpha-\nu} \, g(u) \, R_{\nu-\alpha,\nu}(u,t) \, du, \qquad (2.318)$$

where the unknown function $g(u)$ satisfies the Fredholm integral equation of the second kind

$$g(u) \, u^{\alpha-\nu} + \frac{2u}{\pi} \sin \pi\alpha \int_a^\infty v^{\alpha-\nu} g(v) \left\{ \int_0^a s \, \frac{I_\nu(s)}{K_\nu(s)} \, K_{\nu-\alpha}(vs) \, K_{\alpha-\nu}(us) \, ds \right\} dv$$

$$= \frac{2^{-2\alpha} u^{\nu-\alpha}}{\{\Gamma(1+\alpha)\}^2} \frac{d}{du} \int_u^\infty x^{1-\nu} F_2(x) \, (x^2 - u^2)^\alpha \, dx, \quad a < u < \infty. \qquad (2.319)$$

Equation (2.319) can be reduced to the form

$$g(u) \, u^{\alpha-\nu} + \frac{2u}{\pi} \sin \pi\alpha \int_a^\infty v^{\alpha-\nu} g(v) \, \{M_1(v,u) + M_2(v,u)\} \, dv$$

$$= \frac{2^{-2\alpha} \, u^{\nu-\alpha}}{\{\Gamma(1+\alpha)\}^2} \frac{d}{du} \int_u^\infty x^{1-\nu} f_2(x) \, (x^2 - u^2)^\alpha \, dx, \quad a < u < \infty, \qquad (2.320)$$

where

$$M_1(v,u) = \int_0^\infty s \, \frac{I_\nu(s)}{K_\nu(s)} \, K_{\nu-\alpha}(vs) K_{\nu-\alpha}(us) \, ds, \quad a < v, \ u < \infty,$$

and

$$M_2(v,u) = \frac{\pi \, \text{cosec} \, \pi\alpha}{2^\alpha \Gamma(\alpha+1)} \int_0^\infty t^{2\alpha} \, \frac{\omega(t) \, R_{\nu-\alpha,\nu}(v,t) \, R_{\nu-\alpha,\nu}(u,t)}{J_\nu^2(t) + Y_\nu^2(t)} \, dt.$$

2.4.3 Nasim (1991) generalized the above pairs of dual integral equations discussed in §2.4.1 and §2.4.2. He considered the following dual integral equations

$$W_{\nu-k,\nu}^{-1} \left[t^{-2\alpha} \, f(t); x \right] = f_1(x), \ a \leq x \leq b, \qquad (2.321)$$

$$W_{\nu-k,\nu}^{-1} \left[t^{-2\beta} \, f(t); x \right] = -f_2(x), \ b < x < \infty, \qquad (2.322)$$

where $k = 1, 2, 3, \ldots, \nu > -1$. The notation $W_{\mu,\nu}^{-1}[f(t); x]$ is defined by

$$W_{\mu,\nu}^{-1} \, [f(t); x] = \int_0^\infty \frac{t \, f(t) \, R_{\mu,\nu}(x; t, a)}{J_\nu^2(at) + Y_\nu^2(at)} \, dt, \qquad (2.323)$$

with

$$W_{\mu,\nu}\left[\overline{f}(t); x\right] = \int_a^\infty x\,\overline{f}(x)\,R_{\mu,\nu}(x;t,a)\,dx,\ 0 < t < \infty,\ a > 0, \tag{2.324}$$

where

$$R_{\mu,\nu}(x;t,a) = J_\mu(xt)\,Y_\nu(at) - Y_\mu(xt)\,J_\nu(at). \tag{2.325}$$

Nasim (1991) solved the above pair (2.321) and (2.322) by using the associated Weber−Orr integral transform (cf. Nasim 1989) of the form (see also Appendix A.1.5):

$$\overline{f}(\xi) = W_{\nu-k,\nu}\left[f(x);\xi\right], \tag{2.326}$$

$$f(x) = W_{\nu-k,\nu}^{-1}\left[\overline{f}(\xi);x\right], \tag{2.327}$$

where $0 < k < \frac{\nu}{2} + \frac{3}{4}$, $k = 1, 2, 3, \ldots$ and $x^{k+\frac{1}{2}} f(x)$ is summable in (a,∞).

To solve the above pair, first we decompose it into two sets of dual integral equations for $f_1 = 0$ and $f_2 = 0$ respectively. The solution of the system can then be obtained by adding the solutions of the two sets.

We consider the dual integral equations

$$W_{\nu-k,\nu}^{-1}\left[t^{-2\alpha} f(t); x\right] = 0,\ a \leq x \leq b, \tag{2.328}$$

$$W_{\nu-k,\nu}^{-1}\left[t^{-2\beta} f(t); x\right] = -f_2(x),\ b < x < \infty. \tag{2.329}$$

Case (i): $-1 < \alpha - \beta < 0$.

We set

$$W_{\nu-k,\nu}^{-1}\left[t^{-2\alpha} f(t); x\right] = H(x - b)\,\Psi(x), \tag{2.330}$$

where $H(x)$ is the Heaviside unit function and $\Psi(x)$ is an unknown function. Then due to formula (2.326), for $0 < k < \frac{\nu}{2} + \frac{3}{4}$, (2.328) and (2.330) produce

$$\begin{aligned}
t^{-2\alpha} f(t) &= W_{\nu-k,\nu}[H(x - b)\Psi(x); t] \\
&= \int_b^\infty R_{\nu-k,\nu}(t;x,a)\,x\,\Psi(x)\,dx.
\end{aligned} \tag{2.331}$$

Using the relation (cf. Nasim 1991) for $0 < \beta < \frac{\eta}{2} + \frac{3}{4}$,

$$\int_s^\infty (u^2 - s^2)^{\beta-1} \, u^{1-\eta} \, H_{\nu,\eta,\mu,\alpha}\,(u,t) \, dt$$

$$= 2^{\beta-1} \, \Gamma(\beta) \, s^{\beta-\eta} \, H_{\nu,\eta-\beta,\mu,\alpha-\beta}(s,t), \qquad (2.332)$$

for $-1 < \alpha - \beta < 0$, $\beta - \alpha < \nu - k + \frac{3}{2}$, (2.331) becomes

$$f(t) = \frac{2^{\alpha-\beta+1}}{\Gamma(\beta-\alpha)} \, t^{\beta+\alpha} \int_b^\infty x^{1+\nu-k} \, \Psi(x) \left\{ \int_x^\infty (u^2 - x^2)^{\beta-\alpha-1} \, u^{k-\nu+\alpha-\beta+1} \right.$$

$$\left. \times R_{\nu-k+\beta-\alpha,\nu}(t; u, a) \, du \right.$$

$$= \frac{2^{\alpha-\beta+1}}{\Gamma(\beta-\alpha)} \, t^{\beta+\alpha} \int_b^\infty u^{k-\nu+\alpha-\beta+1} R_{\nu-k+\beta-\alpha,\nu}(t; u, a)$$

$$\times \left\{ \int_b^u x^{1+\nu-k} \, (u^2 - x^2)^{\beta-\alpha-1} \Psi(x) \, dx \right\} du$$

$$= t^{\alpha+\beta} \int_b^\infty R_{\nu-k+\beta-\alpha,\nu}(t; u, a) \, \psi(u) \, du, \qquad (2.333)$$

where

$$\psi(u) = \frac{2^{\alpha-\beta+1}}{\Gamma(\beta-\alpha)} \, u^{k-\nu+\alpha-\beta+1} \int_b^u x^{1+\nu-k} \, (u^2 - x^2)^{\beta-\alpha-1} \, \Psi(x) \, dx,$$

which is an unknown function to be determined. From (2.329), for $x > b$

$$-f_2(x) = W_{\nu-k,\nu}^{-1} \left[t^{-2\beta} f(t); x \right]$$

$$= \int_0^\infty \frac{R_{\nu-k,\nu}(t; x, a)}{J_\nu^2(at) + Y_\nu^2(at)} \, t^{1-2\beta} \, f(t) \, dt.$$

Substituting the expression for $f(t)$ from (2.333) into the above relation and changing the order of integration, we get

$$- f_2(x) = \int_b^\infty \psi(u) \, H_{\nu,\nu-k,\nu-k+\beta-\alpha,\alpha-\beta}(x, u) \, du. \qquad (2.334)$$

Using the relations (A.1.18), (2.291), we have for $b < x < u < \infty$

$$H_{\nu,\nu-k,\nu-k+\beta-\alpha,\alpha-\beta}(x, u) = \frac{2^{1+\alpha-\beta}}{\Gamma(\beta-\alpha)} x^{\nu-k} u^{k-\nu+\alpha-\beta} (u^2 - x^2)^{\beta-\alpha-1} \, H(u-x) + \phi(x, u),$$

78

where

$$\phi(x, u) = \frac{2}{\pi} \sin\left[\pi(k + \alpha - \beta)\right] \int_0^\infty \frac{I_\nu(as)}{K_\nu(as)} K_{\nu-k}(xs) \, K_{\nu-k+\beta-\alpha}(us) \, s^{1+\alpha-\beta} \, ds, \quad (2.335)$$

with $k + 1 > \beta - \alpha$.

Then equation (2.334) becomes

$$-f_2(x) = \frac{2^{1+\alpha-\beta}}{\Gamma(\beta-\alpha)} \, x^{\nu-k} \int_x^\infty u^{k-\nu+\alpha-\beta} \, (u^2 - x^2)^{\beta-\alpha-1} \, \psi(u) \, du$$

$$+ \int_b^\infty \psi(u) \, \phi(x, u) \, du.$$

Now multiplying both sides of the above relation by $2^{\beta-\alpha} x^{1+\alpha-\beta}$ and using the Erdélyi–Kober operator $K_{\eta,\alpha}$, we get

$$K_{\frac{1}{2}(\nu-k+\alpha-\beta+1),\beta-\alpha}[\psi(x)] = -2^{\beta-\alpha} \, x^{1+\alpha-\beta} \int_b^\infty \psi(u) \, \phi(x, u) \, du$$

$$-2^{\beta-\alpha} \, x^{1+\alpha-\beta} \, f_2(x)$$

$$= -F_1(x) - F_2(x), \quad \text{say.} \quad (2.336)$$

Thus, for $-1 < \alpha - \beta < 0$

$$\psi(x) = -K^{-1}_{\frac{1}{2}(\nu-k+\alpha-\beta+1),\beta-\alpha} \left[F_1(x) + F_2(x)\right]$$

$$= -K_{\frac{1}{2}(\nu-k+\beta-\alpha+1),\alpha-\beta} \left[F_1(x) + F_2(x)\right]$$

$$= \frac{1}{2} x^{\nu-k+\beta-\alpha} \frac{d}{dx} \left\{ x^{k-\nu+\alpha-\beta+1} \, K_{\frac{1}{2}(\nu-k+\beta-\alpha-1),1+\alpha-\beta} \left[F_1(x) + F_2(x)\right] \right\}$$

$$= \frac{x^{\nu-k+\beta-\alpha}}{\Gamma(1+\alpha-\beta)} \frac{d}{dx} \int_x^\infty u^{k-\nu+\beta-\alpha} (u^2 - x^2)^{\alpha-\beta} \left[F_1(u) + F_2(u)\right] \, du.$$

Substituting the values of $F_1(u)$ and $F_2(u)$, the above relation becomes

$$\psi(x) = \frac{2^{\beta-\alpha} \, x^{\nu-k+\beta-\alpha}}{\Gamma(1+\alpha-\beta)} \frac{d}{dx} \left[\int_b^\infty \psi(t) \left\{ \int_x^\infty (u^2 - x^2)^{\alpha-\beta} u^{1+k-\nu} \phi(u,t) du \right\} dt \right]$$

$$+ \frac{2^{\beta-\alpha}}{\Gamma(1+\alpha-\beta)} x^{\nu-k+\beta-\alpha} \frac{d}{dx} \left\{ \int_x^\infty u^{1+k-\nu} (u^2 - x^2)^{\alpha-\beta} f_2(u) \, du \right\}$$

$$= I_1 + I_2, \quad \text{say.}$$

(2.337)

Putting the expression for $\phi(u,t)$ into I_1, we get

$$I_1 = \frac{2^{\beta-\alpha+1}}{\pi\Gamma(1+\alpha-\beta)} \, x^{\nu-k+\beta-\alpha} \, \sin\left[\pi(k+\alpha-\beta)\right] \frac{d}{dx} \left[\int_b^\infty \psi(t) \right.$$

$$\times \left\{ \int_x^\infty u^{1+k-\nu} (u^2 - x^2)^{\alpha-\beta} \left(\int_0^\infty \frac{I_\nu(as)}{K_\nu(as)} K_{\nu-k}(us) \, K_{\nu-k+\beta-\alpha}(ts) s^{1+\alpha-\beta} ds \right) du \right\} dt \right].$$

Applying the result (cf. Nasim 1991), for $\beta > 0$

$$\int_s^\infty (u^2 - s^2)^{\beta-1} u^{1-\eta} \, K_\eta(tu) \, du = 2^{\beta-1} \, \Gamma(\beta) \, s^{\beta-\eta} \, t^{-\beta} \, K_{\eta-\beta}(ts),$$

and then simplifying, the above integral becomes

$$I_1 = -\frac{2x}{\pi} \sin\left[\pi(k+\alpha-\beta)\right] \int_b^\infty \psi(t)$$

$$\times \left\{ \int_0^\infty s \, \frac{I_\nu(as)}{K_\nu(as)} \, K_{\nu-k+\beta-\alpha}(xs) \, K_{\nu-k+\beta-\alpha}(ts) \, ds \right\} dt.$$

Then, from (2.337) we obtain, for $-1 < \alpha - \beta < 0$,

$$\psi(x) = -\frac{2x}{\pi} \sin\left[\pi(k+\alpha-\beta)\right] \int_b^\infty \psi(t)$$

$$\times \left\{ \int_0^\infty s \, \frac{I_\nu(as)}{K_\nu(as)} \, K_{\nu-k+\beta-\alpha}(xs) \, K_{\nu-k+\beta-\alpha}(ts) \, ds \right\} dt$$

$$+ \frac{2^{\beta-\alpha} \, x^{\nu-k+\beta-\alpha}}{\Gamma(1+\alpha-\beta)} \frac{d}{dx} \left\{ \int_x^\infty u^{1+k-\nu} (u^2 - x^2)^{\alpha-\beta} f_2(u) \, du \right\}, \quad (2.338)$$

where $0 < k < \frac{\nu}{2} + \frac{3}{4}$, $\beta - \alpha < \nu - k + \frac{3}{2}$ and $\beta - \alpha < k + 1$.

80

Equation (2.338) is a linear Fredholm integral equation of the second kind from which we can find $\psi(t)$ and then the solution of the dual integral equations (2.328) and (2.329) is obtained from (2.333).

Case (ii): $0 < \alpha - \beta < 1$.

The equation (2.328) is satisfied if we assume that

$$W_{\nu-k,\nu}^{-1}\left[t^{-2\alpha}\, f(t); x\right] = H(x - b)\, \Psi(x),$$

where $\Psi(b) = 0$ and $\sqrt{x}\, \Psi(x) \to 0$ as $x \to \infty$. Then, for $0 < k < \frac{\nu}{2} + \frac{3}{4}$,

$$t^{-2\alpha} f(t) = W_{\nu-k,\nu}\left[H(x-b)\, \Psi(x); t\right]$$

$$= \int_b^\infty x\, R_{\nu-k,\nu}(t; x, a)\, \Psi(x)\, dx$$

$$= \frac{1}{t} \int_b^\infty \frac{d}{dx}\left(x^{\nu-k}\, \Psi(x)\right) x^{k-\nu+1}\, R_{\nu-k-1,\nu}(t; x, a)\, dx.$$

Using the relation (2.332), then interchanging the order of integrations we get, for $0 < \alpha - \beta < 1$ and $\beta - \alpha < \nu - k - \frac{1}{2}$

$$f(t) = t^{\alpha+\beta} \int_b^\infty R_{\nu-k+\beta-\alpha,\nu}(t; u, a)\, \psi(u)\, du, \tag{2.339}$$

where

$$\psi(u) = \frac{2^{\alpha-\beta}}{\Gamma(1+\beta-\alpha)}\, u^{k-\nu-\beta+\alpha+1} \int_b^u \frac{d}{dx}\left\{x^{\nu-k}\, \Psi(x)\right\} (u^2 - x^2)^{\beta-\alpha}\, dx.$$

Following a similar analysis as in the previous case, we ultimately obtain a linear Fredholm integral equation of the second kind for the unknown function $\psi(x)$ as

$$\psi(x) = -\frac{2x}{\pi} \sin\left[\pi(k + \alpha - \beta)\right] \int_b^\infty \psi(t)$$

$$\times \left\{\int_0^\infty s\, \frac{I_\nu(as)}{K_\nu(as)}\, K_{\nu-k+\beta-\alpha}(xs)\, K_{\nu-k+\beta-\alpha}(ts)\, ds\right\} dt$$

$$-\frac{2^{\beta-\alpha+1}}{\Gamma(\alpha-\beta)}\, x^{\nu-k+\beta-\alpha+1} \int_x^\infty u^{k-\nu+1}(u^2 - x^2)^{\alpha-\beta-1}\, f_2(u)\, du, \tag{2.340}$$

where $0 < k < \frac{\nu}{2} + \frac{3}{4}$, $\beta - \alpha < \nu - k - \frac{1}{2}$ and $\beta - 1 < k + 1$.

Hence, the solution of the dual integral equations (2.328) and (2.329) is given by

$$f(t) = t^{\alpha+\beta} \int_b^\infty R_{\nu-k+\beta-\alpha,\nu} (t;\, u, a)\, \psi(u)\, du, \qquad (2.341)$$

where $\psi(u)$ satisfies the Fredholm integral equation (2.338) or (2.340) for $-1 < \alpha - \beta < 1$ with $\alpha - \beta \neq 0$.

Next, we consider the pair

$$W_{\nu-k,\nu}^{-1} \left[t^{-2\alpha}\, f(t); x \right] = f_1(x),\ a \leq x \leq b, \qquad (2.342)$$

$$W_{\nu-k,\nu}^{-1} \left[t^{-2\beta}\, f(t); x \right] = 0,\ b < x < \infty. \qquad (2.343)$$

As in previous cases, for $-1 < \alpha - \beta < 0$, the solution of the above pair is

$$\begin{aligned} f(t) &= t^{2\alpha} \int_a^b u\, R_{\nu-k,\nu} (t; u, a)\, f_1(u)\, du \\ &+ t^{\alpha+\beta} \int_b^\infty R_{\nu-k+\beta-\alpha,\nu}(t; u, a)\, \psi(u)\, du, \end{aligned} \qquad (2.344)$$

where $\psi(x)$ satisfies the Fredholm integral equation of the second kind

$$\begin{aligned} \psi(x) &= -\frac{2x}{\pi} \sin\left[\pi(k + \alpha - \beta)\right] \int_b^\infty \psi(t) \\ &\times \left\{ \int_0^\infty s\, \frac{I_\nu(as)}{K_\nu(as)}\, K_{\nu-k+\beta-\alpha}(xs)\, K_{\nu-k+\beta-\alpha}(ts)\, ds \right\} dt \\ &- x \int_a^b t\, H_{\nu,\nu-k+\beta-\alpha,\nu-k,\alpha-\beta}(x, t)\, f_1(t)\, dt, \end{aligned} \qquad (2.345)$$

for $0 < k < \frac{\nu}{2} + \frac{3}{4}$, $\beta - \alpha < \nu - k + \frac{3}{2}$ and $\beta - \alpha < k + 1$.

Similarly, for $0 < \alpha - \beta < 1$, the solution of this pair is

$$\begin{aligned} f(t) &= t^{2\alpha} \int_a^b u\, R_{\nu-k,\nu}(t; u, a)\, f_1(u)\, du \\ &+ t^{\alpha+\beta} \int_b^\infty R_{\nu-k+\beta-\alpha,\nu}(t; u, a)\, \psi(u)\, du, \end{aligned} \qquad (2.346)$$

where $\psi(x)$ satisfies the Fredholm integral equation of the second kind

$$\psi(x) = -\frac{2x}{\pi} \sin\left[\pi(k + \alpha - \beta)\right] \int_b^\infty \psi(t)$$

$$\times \left\{ \int_0^\infty s \, \frac{I_\nu(as)}{K_\nu(as)} \, K_{\nu-k+\beta-\alpha}(xs) \, K_{\nu-k+\beta-\alpha}(ts) \, ds \right\} dt$$

$$-2x \int_0^\infty \frac{R_{\nu-k+\beta-\alpha,\nu}(t; x, a)}{J_\nu^2(at) + Y_\nu^2(at)} \, t^{\alpha-\beta+1} \left\{ \int_a^b u \, R_{\nu-k,\nu}\,(t; u, a) \, f_1(u) \, du \right\} dt,$$
$$(2.347)$$

for $0 < k < \frac{\nu}{2} + \frac{3}{4}$, $\beta - \alpha < \nu - k - \frac{1}{2}$ and $\beta - \alpha < k + 1$.

Hence, the solution of the system (2.342) and (2.343) is given by the relation (2.344) where $\psi(x)$ satisfies the Fredholm integral equation of the second kind (2.345) or (2.347) for $-1 < \alpha - \beta < 1$ with $\alpha - \beta \neq 0$.

Finally, adding the solutions (2.341) and (2.344) we obtain the solution of the pair (2.321) and (2.322).

Special case

Let $\nu = k = 0$, $\alpha = 0$ and $\beta = -\frac{1}{2}$. Then, from (2.328) and (2.329), the dual integral equations

$$W_{0,0}^{-1}\,[f(t); x] = 0, \ a \leq x \leq b,$$
$$W_{0,0}^{-1}\,[t\,f(t); x] = -f_2(x), \ b < x < \infty,$$

have a solution, from (2.339)

$$f(t) = \frac{2}{\pi t} \int_b^\infty \{\cos ut \, Y_0(ua) - \sin ut \, J_0(ua)\} \, u^{-\frac{1}{2}} \, \psi(u) \, du$$

where ψ satisfies the Fredholm integral equation of the second kind, from (2.340)

$$\psi(x) = \sqrt{x} \int_b^\infty t^{-\frac{1}{2}} \, \psi(t) \left\{ \int_0^\infty \frac{I_0(as)}{K_0(as)} \, e^{-s(x+t)} \, ds \right\} dt$$
$$-\sqrt{\frac{2x}{\pi}} \int_x^\infty \frac{x}{\sqrt{u^2 - x^2}} \, f_2(u) \, du,$$

which is the same as (2.282) obtained in §2.4.1.

2.4.4 Now, we study the following dual integral equations (cf. Mandal and Mandal 1996)

$$W_{\nu-\gamma,\nu}^{-1}\left[t^{-2\alpha}\,f(t);x\right] \;=\; f_1(x),\; a \le x \le b, \tag{2.348}$$

$$W_{\nu-\gamma,\nu}^{-1}\left[t^{-2\beta}\,f(t);x\right] \;=\; f_2(x),\; b < x < \infty, \tag{2.349}$$

where $\nu > -\frac{1}{2}$, $\gamma\,(>0)$ is not an integer with $\gamma+\alpha-\beta$ a positive integer. This is a generalization of the pair presented above. This system is reduced to the solution of a Fredholm integral equation of the second kind for $-1 < \alpha-\beta < 1$ with $\alpha-\beta \neq 0$. The method utilized here involves the use of certain multiplying factors to the equations somewhat similar to the method used by Noble (1963) which was presented in §2.1.8.

Multiplying each side of (2.348) by $x^{1+\nu-\gamma}(\xi^2 - x^2)^{\alpha'}$, $\alpha' > -1$ and integrating with respect to x from $a\,(>0)$ to ξ, then interchanging the order of integration on the left side and using the relation (cf. Mandal and Mandal 1996)

$$\int_a^\xi x^{1+\mu}\,(\xi^2 - x^2)^\alpha\,R_{\mu,\nu}(x;t,a)\,dx = 2^\alpha\,\Gamma(\alpha+1)\,t^{\mu+\alpha+1}\,R_{\mu+\alpha+1,\nu}(\xi;t,a)$$

$$-\frac{2^{\mu+1}\,t^{-\alpha-1}}{\pi}\,\xi^{\mu+\alpha+1}\,\Gamma(\mu+1)\,J_\nu(at)\,S_{\alpha-\mu,\alpha+\mu+1}(t\xi)$$

$$-\int_0^a x^{1+\mu}(\xi^2 - x^2)^\alpha R_{\mu,\nu}\,(x;t,a)\,dx, \tag{2.350}$$

for $Re\,\alpha > -1$, $Re\,\mu > -1$, $a > 0$, $S_{\mu,\nu}(z)$ being Lommel's function, we obtain

$$\int_0^\infty \xi^{-2\alpha-\alpha'}\,\frac{R_{\nu-\gamma+\alpha'+1,\gamma}(\xi;t,a)}{J_\nu^2(at)+Y_\nu^2(at)}\,f(t)\,dt$$

$$=\frac{2^{-\alpha'}}{\Gamma(\alpha'+1)}\,\xi^{-\nu+\gamma-\alpha'-1}\int_a^\xi x^{1+\nu-\gamma}(x^2 - \xi^2)^{\alpha'}\,f_1(x)\,dx$$

$$+\frac{2^{\nu-\gamma-\alpha'+1}}{\pi\,\Gamma(\alpha'+1)}\,\Gamma(\nu-\gamma+1)\int_0^\infty u^{-2\alpha-\alpha'}\,\frac{S_{\alpha'-\nu+\gamma,\alpha'+\nu-\gamma+1}(\xi u)}{J_\nu^2(au)+Y_\nu^2(au)}\,J_\nu(au)\,f(u)\,du$$

$$+\frac{2^{-\alpha'}}{\Gamma(\alpha'+1)}\xi^{-\nu+\gamma-\alpha'+1}\int_0^\infty \frac{u^{-2\alpha+1}f(u)}{J_\nu^2(au)+Y_\nu^2(au)}$$

$$\times \left\{ \int_0^a \rho^{1+\nu-\gamma} \, (\xi^2 - \rho^2)^{\alpha'} \, R_{\nu-\gamma,\nu} \, (\rho; u, a) \, d\rho \right\} du, \qquad (2.351)$$

for $a \leq t \leq b$, $\alpha' > -1$, $\nu > \gamma - 1$.

Similarly, multiplying each side of (2.349) by $x^{1-\nu+\gamma}(x^2 - \xi^2)^{\beta'}$, $\beta' > -1$ and integrating with respect to x from ξ to ∞, then interchanging the order of integration and using the relation (cf. Mandal and Mandal 1996)

$$\int_\xi^\infty x^{1-\mu} \, R_{\mu,\nu}(x; t, a)(x^2 - \xi^2)^\beta \, dx = 2^\beta \Gamma(\beta+1) \, \xi^{1+\beta-\mu} \, t^{-\beta-1} \, R_{\mu-\beta-1,\nu} \, (\xi; t, a), \quad (2.352)$$

for $Re(\beta) > -1$, $Re(\mu - 2\beta) > \frac{1}{2}$, we get

$$\int_0^\infty t^{-2\beta-\beta'} \, \frac{R_{\nu-\gamma-\beta'-1,\nu}(\xi; t, a)}{J_\nu^2(at) + Y_\nu^2(at)} \, f(t) \, dt$$

$$= \frac{2^{-\beta'}}{\Gamma(\beta'+1)} \xi^{\nu-\gamma-\beta'-1} \int_\xi^\infty x^{1-\nu+\gamma}(x^2 - \xi^2)^{\beta'} \, f_2(x) \, dx, \qquad (2.353)$$

for $b < \xi < \infty$, $-1 < \beta' < \frac{1}{2}(\nu - \gamma - \frac{1}{2})$.

In relations (2.351) and (2.353), we should like to set

$$\nu - \gamma + \alpha' + 1 = \nu - \gamma - \beta' + 1 \quad \text{and} \quad -2\alpha - \alpha' = -2\beta - \beta'$$

so that the orders of $R_{\mu,\nu} \, (\xi; t, a)$ and the powers of t would be identical. This would mean, on solving these equations, that

$$\alpha' = -1 - (\alpha - \beta), \ \beta' = -1 + (\alpha - \beta).$$

Since α' and β' must be greater than -1, this is impossible. To overcome this difficulty, we consider the following two cases separately.

Case I: $-1 < \alpha - \beta < 0$, $\alpha - \beta < \min\left\{\frac{1}{2}\left(\nu - \gamma - \frac{1}{2}\right), \frac{1}{2}\left(\nu - 2\gamma + \frac{3}{2}\right)\right\}$.

We take $\alpha' = -1 - (\alpha - \beta)$, $\beta' = \alpha - \beta$. Because of the conditions on α', β' in (2.351) and (2.353), we must have $-1 < \alpha - \beta < 0$, $\alpha - \beta < \frac{1}{2}(\nu - \gamma - \frac{1}{2})$. Then the above two equations give

$$\int_0^\infty t^{1-\alpha-\beta}\, \frac{R_{\nu-\gamma-\alpha+\beta,\nu}(\xi;t,a)}{J_\nu^2(at)+Y_\nu^2(at)}\, f(t)\, dt$$

$$= \frac{2^{1+\alpha-\beta}}{\Gamma(\beta-\alpha)}\xi^{-\nu+\gamma+\alpha-\beta}\int_a^t x^{1+\nu-\gamma}(\xi^2-x^2)^{\beta-\alpha-1}\, f_1(x)\, dx$$

$$+ \frac{2^{\nu-\gamma+\alpha-\beta+2}\Gamma(\nu-\gamma+1)}{\pi\Gamma(\beta-\alpha)}\int_0^\infty \frac{S_{-\nu+\gamma-\alpha+\beta-1,\nu-\gamma-\alpha+\beta}(\xi u)}{u^{\alpha+\beta-1}\{J_\nu^2(au)+Y_\nu^2(au)\}}\, J_\nu(au)\, f(u)du$$

$$+ \frac{2^{1+\alpha-\beta}}{\Gamma(\beta-\alpha)}\,\xi^{-\nu+\gamma+\alpha-\beta}\int_0^\infty \frac{u^{-2\alpha+1}f(u)}{J_\nu^2(au)+Y_\nu^2(au)}$$

$$\times \left\{\int_0^a \rho^{1+\nu-\gamma}(\xi^2-\rho^2)^{\beta-\alpha-1}\, R_{\nu-\gamma,\nu}(\rho;\, u,a)\, d\rho\right\} du, \tag{2.354}$$

for $a \leq \xi \leq b$, and

$$\int_0^\infty t^{-\alpha-\beta}\, \frac{R_{\nu-\gamma-\alpha+\beta-1,\nu}(\xi;t,a)}{J_\nu^2(at)+Y_\nu^2(at)}f(t)\, dt$$

$$= \frac{2^{\beta-\alpha}}{\Gamma(1+\alpha-\beta)}\xi^{\nu-\gamma-\alpha+\beta-1}\int_\xi^\infty x^{1-\nu+\gamma}\,(x^2-\xi^2)^{\alpha-\beta}\, f_2(x)\, dx,\ b < \xi < \infty. \tag{2.355}$$

Now multiply both sides of (2.355) by $\xi^{-\nu+\gamma+\alpha-\beta+1}$ and differentiate with respect to ξ, then using the relation

$$\frac{1}{\xi}\, \frac{d}{d\xi}\, [\xi^{-\mu}\, R_{\mu,\nu}\, (\xi;t,a)] = -\xi^{-\mu-1}\, R_{\mu+1,\nu}(\xi;t,a),$$

we obtain

$$\int_0^\infty t^{1-\alpha-\beta}\, \frac{R_{\nu-\gamma-\alpha+\beta,\nu}(\xi;t,a)}{J_\nu^2(at)+Y_\nu^2(at)}f(t)\, dt$$

$$= -\frac{2^{\beta-\alpha}\xi^{\nu-\gamma-\alpha+\beta-1}}{\Gamma(1+\alpha-\beta)}\frac{d}{d\xi}\int_\xi^\infty x^{1-\nu+\gamma}(x^2-\xi^2)^{\alpha-\beta}\, f_2(x)\, dx,\ b < \xi < \infty. \tag{2.356}$$

Equations (2.354) and (2.356) have identically the same left-hand side which is therefore defined for all $\xi \geq a$. Since we have assumed $\gamma + \alpha - \beta$ to be a positive integer and $\alpha-\beta < \frac{1}{2}(\nu-2\gamma+\frac{3}{2})$, therefore, by the associated Weber$-$Orr transform formula (2.326), we obtain

$$f(t) = t^{\alpha+\beta} \left\{ \frac{2^{1+\alpha-\beta}}{\Gamma(\beta-\alpha)} \int_a^b \xi^{1-\nu+\gamma+\alpha-\beta} R_{\nu-\gamma-\alpha+\beta,\nu}(\xi;t,a) \; F_1(\xi) \; d\xi \right.$$

$$+ \frac{2^{\beta-\alpha}}{\Gamma(1+\alpha-\beta)} \int_b^\infty \xi^{\nu-\gamma-\alpha+\beta} R_{\nu-\gamma-\alpha+\beta,\nu}(\xi;t,a) \; F_2(\xi) \; d\xi$$

$$\left. + \frac{2^{1+\alpha-\beta}}{\Gamma(\beta-\alpha)} \int_0^\infty u^{1-\alpha-\beta} \; K(u,t) \; f(u) \; du \right\}, \qquad (2.357)$$

where

$$F_1(\xi) = \int_a^\xi x^{1+\nu-\gamma}(\xi^2-x^2)^{\beta-\alpha-1} \; f_1(x) \; dx$$

$$F_2(\xi) = -\frac{d}{d\xi} \int_\xi^\infty x^{1-\nu+\gamma}(x^2-\xi^2)^{\alpha-\beta} \; f_2(x) \; dx,$$

$$K(u,t) = \frac{\frac{1}{\pi}2^{1+\nu-\gamma}\Gamma(1+\nu-\gamma) \; J_\nu(au) \; G_1(u,t) + u^{-\alpha+\beta}G_2(u,t)}{J_\nu^2(au) + Y_\nu^2(au)},$$

$$G_1(u,t) = \int_a^b \xi \; S_{-\nu+\gamma-\alpha+\beta-1,\nu-\gamma-\alpha+\beta}(\xi u) \; R_{\nu-\gamma-\alpha+\beta,\nu}(\xi;t,a) \; d\xi,$$

$$G_2(u,t) = \int_a^b \xi^{1-\nu+\gamma+\alpha-\beta} \; K_1(u,\xi) \; R_{\nu-\gamma-\alpha+\beta,\nu} \; (\xi;t,a) \; d\xi$$

and

$$K_1(u,\xi) = \int_0^a \rho^{1+\nu-\gamma}(\xi^2-\rho^2)^{\beta-\alpha-1} \; R_{\nu-\gamma,\nu} \; (\rho;u,a) \; d\rho.$$

Equation (2.357) is a Fredholm integral equation of the second kind.

Case II: $0 < \alpha - \beta < 1$, $\alpha - \beta < \frac{1}{2}(\nu - \gamma + \frac{3}{2})$.

In this case, we choose $\alpha' = \beta - \alpha$, $\beta' = -1 + \alpha - \beta$. Because of the conditions on α', β' in (2.351) and (2.353) we must have $0 < \alpha - \beta < 1$, $\alpha - \beta < \frac{1}{2}(\nu - \gamma + \frac{3}{2})$. Then (2.351) and (2.353) become

$$\int_0^\infty t^{-\alpha-\beta} \frac{R_{\nu-\gamma-\alpha+\beta+1,\nu}(\xi;t,a)}{J_\nu^2(at) + Y_\nu^2(at)} f(t) \, dt$$

$$= \frac{2^{\alpha-\beta}}{\Gamma(1+\beta-\alpha)} \xi^{-\nu+\gamma+\alpha-\beta-1} \int_a^\xi x^{1+\nu-\gamma}(\xi^2 - x^2)^{\beta-\alpha} \, f_1(x) \, dx$$

$$+ \frac{2^{\nu-\gamma+\alpha-\beta+1}\Gamma(\nu-\gamma+1)}{\pi \, \Gamma(\beta-\alpha+1)} \int_0^\infty \frac{S_{-\nu+\gamma-\alpha+\beta,\nu-\gamma-\alpha+\beta+1}(\xi u)}{u^{\alpha+\beta} \{J_\nu^2(au) + Y_\nu^2(au)\}} \, J_\nu(au) \, f(u) \, du$$

$$+ \frac{2^{\alpha-\beta}}{\Gamma(1+\beta-\alpha)} \, \xi^{-\nu+\gamma+\alpha-\beta-1} \int_0^\infty \frac{u^{-2\alpha+1} f(u)}{J_\nu^2(au) + Y_\nu^2(au)}$$

$$\times \left\{ \int_0^a \rho^{1+\nu-\gamma}(\xi^2 - \rho^2)^{\beta-\alpha} \, R_{\nu-\gamma,\nu}(\rho; \, u, a) \, d\rho \right\} du, \quad a \le \xi \le b, \qquad (2.358)$$

and

$$\int_0^\infty t^{1-\alpha-\beta} \frac{R_{\nu-\gamma-\alpha+\beta,\nu}(\xi;t,a)}{J_\nu^2(at) + Y_\nu^2(at)} f(t) \, dt$$

$$= \frac{2^{1-\alpha+\beta}}{\Gamma(\alpha-\beta)} \xi^{\nu-\gamma-\alpha+\beta} \int_\xi^\infty x^{1-\nu+\gamma} \, (x^2 - \xi^2)^{\alpha-\beta-1} \, f_2(x) \, dx, \quad b < \xi < \infty. \qquad (2.359)$$

Now we multiply both sides of (2.358) by $\xi^{\nu-\gamma-\alpha+\beta+1}$ and differentiate with respect to ξ and then using the relation

$$\frac{1}{\xi} \frac{d}{d\xi} \left[\xi^\mu \, R_{\mu,\nu} \, (\xi;t,a) \right] = \xi^{\mu-1} \, t \, R_{\mu-1,\nu}(\xi;t,a),$$

we obtain

$$\int_0^\infty t^{1-\alpha-\beta} \frac{R_{\nu-\gamma-\alpha+\beta,\nu}(\xi;t,a)}{J_\nu^2(at) + Y_\nu^2(at)} \, f(t) \, dt$$

$$= \frac{2^{\alpha-\beta}}{\Gamma(1+\beta-\alpha)} \xi^{-\nu+\gamma+\alpha-\beta-1} \frac{d}{d\xi} \int_a^\xi x^{1+\nu-\gamma}(\xi^2 - x^2)^{\beta-\alpha} \, f_1(x) \, dx$$

$$+ \frac{2^{\nu-\gamma+\alpha-\beta+1}}{\pi\Gamma(1+\beta-\alpha)}\Gamma(\nu-\gamma+1) \, \xi^{-\nu+\gamma+\alpha-\beta-1} \frac{d}{d\xi} \left\{ \xi^{\nu-\gamma-\alpha+\beta+1} \right.$$

$$\times \int_0^\infty u^{-\alpha-\beta} \frac{S_{-\nu+\gamma-\alpha+\beta,\nu-\gamma-\alpha+\beta+1}(\xi u)}{J_\nu^2(au) + Y_\nu^2(au)} \, J_\nu(au) \, f(u) \, du \Bigg\}$$

88

$$+\frac{2^{\alpha-\beta}}{\Gamma(1+\beta-\alpha)}\ \xi^{-\nu+\gamma+\alpha-\beta-1}\ \frac{d}{d\xi}\ \int_0^\infty \frac{u^{-2\alpha+1}f(u)}{J_\nu^2(au)+Y_\nu^2(au)}$$

$$\times \left\{ \int_0^a \rho^{1+\nu-\gamma}(\xi^2-\rho^2)^{\beta-\alpha}\ R_{\nu-\gamma,\nu}(\rho;\ u,a)\ d\rho \right\}\ du,\quad a\le\xi\le b. \qquad (2.360)$$

As before, equations (2.359) and (2.360) have identically the same left-hand side which is therefore defined for all $\xi\ge a$. Since we have assumed $\gamma+\alpha-\beta$ to be a positive integer and $\alpha-\beta<\frac{1}{2}(\nu-\gamma+\frac{3}{2})$, therefore by the associated Weber $-$Orr integral transform formula (2.326), we find that

$$f(t)\ =\ t^{\alpha+\beta}\left\{\frac{2^{\alpha-\beta}}{\Gamma(1-\alpha+\beta)}\int_a^b \xi^{-\nu+\gamma+\alpha-\beta}\ R_{\nu-\gamma-\alpha+\beta,\nu}\ (\xi;t,a)\ \overline{F}_1(\xi)\ d\xi\right.$$

$$+\frac{2^{1-\alpha+\beta}}{\Gamma(\alpha-\beta)}\int_b^\infty \xi^{1+\nu-\gamma+\alpha-\beta}\ R_{\nu-\gamma-\alpha+\beta,\nu}\ (\xi;t,a)\ \overline{F}_2(\xi)\ d\xi$$

$$+\frac{2^{\alpha-\beta}}{\Gamma(1-\alpha+\beta)}\int_0^\infty u^{-\alpha-\beta}\ \overline{K}(u,t)\ f(u)\ du\Bigg\}, \qquad (2.361)$$

where

$$\overline{F}_1(\xi)\ =\ \frac{d}{d\xi}\int_a^\xi x^{1+\nu-\gamma}\ (\xi^2-x^2)^{\beta-\alpha}\ f_1(x)\ dx,$$

$$\overline{F}_2(\xi)\ =\ \int_\xi^\infty x^{1-\nu+\gamma}(x^2-\xi^2)^{\alpha-\beta-1}\ f_2(x)\ dx.$$

$$\overline{K}(u,t)\ =\ \frac{\frac{1}{\pi}2^{1+\nu-\gamma}\Gamma(1+\nu-\gamma)\ J_\nu(au)\ \overline{G}_1(u,t)+u^{1-\alpha+\beta}\ \overline{G}_2(u,t)}{J_\nu^2(au)+Y_\nu^2(au)},$$

$$\overline{G}_1(u,t)\ =\ \int_a^b \xi^{-\nu+\gamma+\alpha-\beta}\ \frac{d}{d\xi}\left\{\xi^{\nu-\gamma-\alpha+\beta-1}\ S_{-\nu+\gamma-\alpha+\beta,\nu-\gamma-\alpha+\beta+1}(\xi u)\right\}$$

$$\times R_{\nu-\gamma+\beta-\alpha,\nu}\ (\xi;t,a)\ d\xi,$$

$$\overline{G}_2(u,t)\ =\ \int_a^b \xi^{-\nu+\gamma+\alpha-\beta}\ \overline{K}_1(u,\xi)\ R_{\nu-\gamma+\beta-\alpha,\nu}\ (\xi;t,a)\ d\xi,$$

and

$$\overline{K}_1(u,\xi)=\frac{d}{d\xi}\int_0^a \rho^{1+\nu-\gamma}(\xi^2-\rho^2)^{\beta-\alpha}\ R_{\nu-\gamma,\nu}\ (\rho;u,a)\ d\rho.$$

Equation (2.361) is a Fredholm integral equation of the second kind.

Special case

If we make $a \to 0$ and $\nu > 0$ with $\lim\limits_{a \to 0} \dfrac{f(t)}{Y_\nu(at)} = A(t)$ and α replaced by $\frac{1-\alpha}{2}$ and β by $\frac{1}{2}$ and putting $\nu - \gamma = \mu$, the dual integral equations (2.348) and (2.349) become

$$\int_0^\infty t^\alpha A(t) J_\mu(xt) \, dt = f_1(x), \ a \le x \le b,$$

$$\int_0^\infty A(t) J_\mu(xt) \, dt = f_2(x), \ x > b.$$

This pair was considered by Noble (1963). From relations (2.357) and (2.361), the solution of this pair becomes:

For $0 < \alpha < 2, \ \mu > \frac{1}{2} - \alpha$,

$$A(t) = \frac{(2t)^{1-\frac{\alpha}{2}}}{\Gamma(\frac{\alpha}{2})} \int_0^b \xi^{-\mu-\frac{\alpha}{2}+1} J_{\mu+\frac{\alpha}{2}} (t\xi) F_1(\xi) \, d\xi$$

$$+ \frac{2^{\frac{\alpha}{2}} t^{1-\frac{\alpha}{2}}}{\Gamma(1-\frac{\alpha}{2})} \int_b^\infty \xi^{\mu+\frac{\alpha}{2}} J_{\mu+\frac{\alpha}{2}} (t\xi) F_2(\xi) \, d\xi,$$

where

$$F_1(\xi) = \int_0^\infty x^{1+\mu} f_1(x) (\xi^2 - x^2)^{-1+\frac{\alpha}{2}} \, dx,$$

$$F_2(\xi) = -\frac{d}{d\xi} \int_\xi^\infty x^{1-\mu} f_2(x) (x^2 - \xi^2)^{-\frac{\alpha}{2}} \, dx,$$

and for $-2 < \alpha < 0, \ \mu > -\frac{3}{2} - \alpha$,

$$A(t) = \frac{2^{-\frac{\alpha}{2}} t^{-\frac{\alpha}{2}}}{\Gamma(1+\frac{\alpha}{2})} \int_0^b \xi^{-\mu-\frac{\alpha}{2}} J_{\mu+\frac{\alpha}{2}} (t\xi) \overline{F}_1(\xi) \, d\xi$$

$$+ \frac{2^{1+\frac{\alpha}{2}} t^{1-\frac{\alpha}{2}}}{\Gamma(-\frac{\alpha}{2})} \int_b^\infty \xi^{\mu+\frac{\alpha}{2}+1} J_{\mu+\frac{\alpha}{2}} (t\xi) \overline{F}_2(\xi) \, d\xi,$$

where

$$\overline{F}_1(\xi) = \frac{d}{d\xi} \int_0^\xi x^{1+\mu} f_1(x) (\xi^2 - x^2)^{\frac{\alpha}{2}} \, dx,$$

$$\overline{F}_2(\xi) = \int_\xi^\infty x^{1-\mu} f_2(x) (x^2 - \xi^2)^{-1-\frac{\alpha}{2}} \, dx.$$

This result was obtained by Noble (1963) directly.

90

Chapter 3

Dual integral equations with spherical harmonic kernel

Dual integral equations with Legendre function $P_{-\frac{1}{2}+i\tau}(\cosh x)$ as kernel occur in mixed boundary value problems involving conical or spherical regions. Babloian (1964) first handled these types of dual integral equations. He solved these pairs of integral equations by the use of the integral representations of Legendre functions and the classical Mehler–Fock integral transform. The works of Babloian are discussed in some detail in the book by Sneddon (1972) and also mentioned in the book of Virchenko (1989), p. 79.

In this chapter, we present methods of solutions to some dual integral equations involving spherical harmonics, e.g., Legendre, associated Legendre and generalized associated Legendre functions as kernel together with their applications to some physical problems of mathematical physics in three sections.

3.1 Kernels involving Legendre functions

In this section, we first consider a general pair of integral equations (cf. Lebedev and Skal'skaya 1969) of the form

$$\int_0^\infty f(\tau)\, P_{-\frac{1}{2}+i\tau}(\cosh x)\, d\tau \;=\; g(x),\; 0 \le x < a, \tag{3.1}$$

$$\int_0^\infty f(\tau)\, \omega_\mu(\tau)\, P_{-\frac{1}{2}+i\tau}(\cosh x)\, d\tau \;=\; 0,\; a < x < \infty, \tag{3.2}$$

where

$$\omega_\mu(\tau) = 4\pi^2 \left[\cosh \pi\tau \ \Gamma\left(\frac{1}{4} + \frac{\mu}{2} + \frac{i\tau}{2}\right) \ \Gamma\left(\frac{1}{4} + \frac{\mu}{2} - \frac{i\tau}{2}\right) \right.$$

$$\left. \times \Gamma\left(\frac{1}{4} - \frac{\mu}{2} + \frac{i\tau}{2}\right) \ \Gamma\left(\frac{1}{4} - \frac{\mu}{2} - \frac{i\tau}{2}\right) \right]^{-1}, \qquad (3.3)$$

with μ as an arbitrary real parameter. We can assume that $\mu \geq 0$ as $\omega_{-\mu}(\tau) = \omega_\mu(\tau)$. For $\mu = 0$, the dual integral equations (3.1) and (3.2) correspond to the pair

$$\left. \begin{array}{rl} \displaystyle\int_0^\infty f(\tau) \ P_{-\frac{1}{2}+i\tau}(\cosh x) \ d\tau &= \ g(x), \ 0 \leq x < a, \\[3mm] \displaystyle\int_0^\infty f(\tau) \ \frac{\cosh \pi\tau}{\pi[P_{-\frac{1}{2}+i\tau}(0)]^2} \ P_{\frac{1}{2}+i\tau}(\cosh x) \ d\tau &= \ 0, \ a < x < \infty. \end{array} \right\} \qquad (3.4)$$

This pair arises in the solution of a boundary value problem with mixed boundary conditions prescribed on the surface of a one-sheet hyperboloid of revolution (cf. Lebedev and Skal'skaya 1969). For $\mu = \frac{1}{2}$, the pair reduces to

$$\left. \begin{array}{rl} \displaystyle\int_0^\infty f(\tau) \ P_{-\frac{1}{2}+i\tau}(\cosh x) \ d\tau &= \ g(x), \ 0 \leq x < a, \\[3mm] \displaystyle\int_0^\infty \tau \ \tanh \pi\tau \ f(\tau) \ P_{-\frac{1}{2}+i\tau}(\cosh x) \ d\tau &= \ 0, \ a < x < \infty. \end{array} \right\} \qquad (3.5)$$

This pair has various applications in the solution of contact problems in the theory of elasticity, electrostatics, etc., involving conical boundaries (cf. Babloian 1964).

The solution of the pair (3.1) and (3.2) can be obtained explicitly. To find the solution of these equations, we first assume that $0 \leq \mu \leq \frac{1}{2}$ and $g(x)$ is a known continuously differentiable function.

We consider the solution of the pair (3.1) and (3.2) in the form

$$f(\tau) = \frac{2\tau \ \tanh \pi\tau}{\pi \ \omega_\mu(\tau)} \int_0^a F\left(\frac{1}{4} + \frac{\mu}{2} + \frac{i\tau}{2}, \ \frac{1}{4} + \frac{\mu}{2} - \frac{i\tau}{2}; \frac{1}{2}; -\sinh^2 t\right) \cosh t \ \phi(t) \ dt, \qquad (3.6)$$

where $\phi(t)$ is an unknown function to be determined with the assumption that it is continuous together with its first derivative in $(0, a)$.

Integrating the right side of (3.6) by parts and then using the known relation

$$F\left(a, b; \ \frac{1}{2}; -z^2\right) = \frac{d}{dz}\left\{ z \ F\left(a, b; \ \frac{3}{2}; -z^2\right) \right\},$$

92

we obtain

$$f(\tau) = \frac{2\tau \tanh \pi\tau}{\pi \omega_\mu(\tau)} \left\{ \phi(a) \sinh a \ F\left(\frac{1}{4} + \frac{\mu}{2} + \frac{i\tau}{2}, \frac{1}{4} + \frac{\mu}{2} - \frac{i\tau}{2}; \frac{3}{2}; -\sinh^2 a\right) \right.$$

$$\left. - \int_0^a F\left(\frac{1}{4} + \frac{\mu}{2} + \frac{i\tau}{2}, \frac{1}{4} + \frac{\mu}{2} - \frac{i\tau}{2}; \frac{3}{2}; -\sinh^2 t\right) \cosh t \ \phi'(t) \ dt \right\}. \quad (3.7)$$

Substituting the above expression of $f(\tau)$ in (3.2) and then using the result (A.2.6), it is observed that equation (3.2) is satisfied identically. Now, substituting the value of $f(\tau)$ from (3.6) into (3.1) and then utilizing the result (A.2.5), we find an integral equation of the first kind for the unknown function $\phi(t)$ as given by

$$\int_0^x F\left(-\frac{\mu}{2}, \frac{\mu}{2}; \frac{1}{2}; \frac{\cosh^2 t - \cosh^2 x}{\cosh^2 t}\right) \frac{\cosh^{1-\mu} t \ \phi(t)}{\sqrt{\cosh^2 x - \cosh^2 t}} dt = \frac{\pi}{2} \ g(x), \ 0 \le x < a. \quad (3.8)$$

Putting $u = \cosh^2 x$, $v = \cosh^2 t$, $\psi(v) = \frac{\sqrt{\pi} \ v^{-\frac{\mu}{2}}}{\sqrt{v-1}} \ \phi\left(\cosh^{-1} \sqrt{v}\right)$, $h(u) = \frac{\pi}{2} \ g(\cosh^{-1} \sqrt{u})$, into the above equation (3.8), it reduces to the simple form

$$\int_1^u F\left(-\frac{\mu}{2}, \frac{\mu}{2}; \frac{1}{2}; 1 - \frac{u}{v}\right) \frac{\psi(v)}{\sqrt{\pi(u-v)}} = h(u), \ u \ge 1. \quad (3.9)$$

The above integral equation belongs to a class of integral equations investigated by Love (1967). This class of integral equations includes many equations of practical importance, in particular Abel's integral equation which plays an important role in the solution of dual integral equations (3.4) and (3.5).

Using the theory developed by Love (1967), the solution of the integral equation (3.9) can be obtained as

$$\psi(v) = v^{\frac{\mu}{2}} \frac{d}{dv} \left\{ v^{\frac{\mu}{2}} \int_1^v F\left(\frac{\mu}{2}, 1 - \frac{\mu}{2}; \frac{1}{2}; 1 - \frac{u}{v}\right) \frac{g(u)}{\sqrt{\pi(v-u)}} \ du \right\}. \quad (3.10)$$

Substituting back $u = \cosh^2 x$, $v = \cosh^2 t$, the above equation produces

$$\phi(t) = (\cosh t)^{2\mu-1} \frac{d}{dt} \left\{ \cosh^{-\mu} t \int_0^t F\left(\frac{\mu}{2}, 1 - \frac{\mu}{2}; \frac{1}{2}; 1 - \frac{\cosh^2 x}{\cosh^2 t}\right) \right.$$

$$\left. \times \frac{\cosh x \ \sinh x \ g(x)}{\sqrt{\cosh^2 t - \cosh^2 x}} \ dx \right\}. \quad (3.11)$$

93

Therefore, we obtain the explicit solution to the dual integral equations (3.1) and (3.2) by (3.6) where $\phi(t)$ is given by (3.11).

The method presented above for the solution of the pair (3.1) and (3.2) can be generalized to the case $\mu > \frac{1}{2}$, but the details of the calculations will be somewhat cumbersome. Without presenting the explicit form of the solution and details of calculations, a brief outline of this method is now given when $0 \leq \mu < \infty$.

As discussed above, the solution of the pair of integral equations can be represented in the form (3.6) for any $\mu > 0$. Then using the result (A.2.6) and equation (3.6), the equation (3.2) is satisfied identically. Substituting (3.6) into the inhomogeneous equation (3.1) and utilizing the result for the integral in (A.2.5) for $\mu > \frac{1}{2}$, we get an integral equation of the first kind for the unknown function $\phi(t)$ as

$$\int_0^x F\left(-\frac{\mu}{2}, \frac{\mu}{2}; \frac{1}{2}; \frac{\cosh^2 t - \cosh^2 x}{\cosh^2 t}\right) \frac{\cosh^{1-\mu} t \; \phi(t)}{\sqrt{\cosh^2 x - \cosh^2 t}} \, dt$$

$$= \frac{\pi}{2} \, g(x) + \sum_{n=0}^{N} \alpha_n \, P_{2n-\mu}(\cosh x), \; 0 \leq x < a,$$

where $N = \left[\frac{1}{2}\left(\mu - \frac{1}{2}\right)\right]$, and the unknown coefficients α_n represent integrals of products of the unknown function $\phi(t)$ with some known functions of the variable t on $(0, a)$. In particular, for $0 \leq \mu \leq \frac{1}{2}$, the sum becomes empty and the additional term is absent in (3.8). Then using the solution (3.11), the determination of the unknown coefficients α_n is reduced to the solution of a system of linear algebraic equations.

As an example of the above problem, for $\mu > \frac{1}{2}$, we consider the dual integral equations

$$\int_0^\infty f(\tau) \, P_{-\frac{1}{2}+i\tau}(\cosh x) \, d\tau = g(x), \; 0 \leq x < a, \tag{3.12}$$

$$\int_0^\infty \omega_1(\tau) \, f(\tau) \, P_{-\frac{1}{2}+i\tau}(\cosh x) \, d\tau = 0, \; a < x < \infty, \tag{3.13}$$

which correspond to a value of $\mu = 1$ and this pair occurs in some physical problems.

In this case, the relation (A.2.5) becomes (cf. Lebedev and Skal'skaya 1969)

$$\cosh t \int_0^\infty \frac{\tau \, \tanh \pi \tau}{\omega_1(\tau)} F\left(\frac{3}{4} + \frac{i\tau}{2}, \frac{3}{4} - \frac{i\tau}{2}; \frac{1}{2}; -\sinh^2 t\right) P_{-\frac{1}{2}+i\tau}(\cosh x) \, d\tau$$

$$= \begin{cases} \dfrac{\cosh x}{\cosh t \sqrt{\cosh^2 x - \cosh^2 t}} - \operatorname{sech} t, & t < x, \\ -\operatorname{sech} t, & t > x. \end{cases} \tag{3.14}$$

We represnt the solution of the dual integral equations (3.12) and (3.13) in the form

$$f(\tau) = \frac{2\tau \, \tanh \pi \tau}{\pi \, \omega_1(\tau)} \int_0^a F\left(\frac{3}{4} + \frac{i\tau}{2}, \frac{3}{4} - \frac{i\tau}{2}; \frac{1}{2}; -\sinh^2 t\right) \cosh t \, \phi(t) \, dt; \tag{3.15}$$

then (3.13) is satisfied identically and thus the equations (3.12) and (3.15) produce an integral equation as given by

$$\cosh x \int_0^x \frac{\phi(t) \, dt}{\cosh t \, \sqrt{\cosh^2 x - \cosh^2 t}} = \frac{\pi}{2} \, g(x) + C, \tag{3.16}$$

where C is given by

$$C = \int_0^a \frac{\phi(t)}{\cosh t} \, dt. \tag{3.17}$$

Solving this integral equation by the method presented above, we obtain

$$\phi(t) = \cosh t \, \frac{d}{dt} \int_0^t \frac{\sinh x \, g(x)}{\sqrt{\cosh^2 t - \cosh^2 x}} \, dx + \frac{2C}{\pi}. \tag{3.18}$$

Multiplying the equation (3.18) by $(\cosh t)^{-1}$ and then integrating from 0 to a, the unknown constant C is obtained as

$$C = \frac{\pi}{\pi - 2\tan^{-1}(\sinh a)} \int_0^a \frac{\sinh x \, g(x)}{\sqrt{\cosh^2 a - \cosh^2 x}} \, dx. \tag{3.19}$$

3.1.1 Next, we consider a pair of integral equations

$$\int_0^\infty \tau^{-1} \, f(\tau) \, \tanh c\tau \, P_{-\frac{1}{2}+i\tau}(\cosh x) \, d\tau = g(x), \quad 0 < x < a, \tag{3.20}$$

$$\int_0^\infty f(\tau) \, P_{-\frac{1}{2}+i\tau}(\cosh x) \, d\tau = 0, \quad a < x < \infty, \tag{3.21}$$

where c is a real positive constant. This pair is somewhat different from the pair discussed above. When $c = \frac{\pi}{2}$, the dual integral equations (3.20) and (3.21) occur in the solution of the electrostatic problem in which a disc charged to a prescribed potential is placed on the flat part of an earthed hollow hemisphere (cf. Dhaliwal and Sing 1987). Several researchers have studied this class of dual integral equations and they reduced the pair to the solution of a Fredholm integral equation of the second kind. An explicit solution to the more general form of the dual integral equations (3.20) and (3.21) is obtained here by using the finite Hilbert transform technique (cf. Dhaliwal and Sing 1987).

To find the solution of this pair, we first multiply (3.20) by $\sinh x/(\cosh t - \cosh x)^{\frac{1}{2}}$ and integrate it with respect to x from 0 to t. Then differentiating both sides with respect to t and using the result

$$\frac{d}{dt} \int_0^t \frac{P_{-\frac{1}{2}+i\tau}(\cosh x) \sinh x}{\sqrt{\cosh t - \cosh x}} \, dx = \sqrt{2} \, \cos \tau t,$$

equation (3.20) then reduces to

$$\int_0^\infty \tau^{-1} f(\tau) \, \cos \tau t \, \tanh c\tau \, d\tau = g_1(t), \ 0 < t < x, \qquad (3.22)$$

where

$$g_1(t) = \frac{1}{\sqrt{2}} \frac{d}{dt} \int_0^t \frac{\sinh x \, g(x)}{\sqrt{\cosh t - \cosh x}} \, dx.$$

Let us now put

$$f(\tau) = \int_0^a \phi(s) \, \cos \tau s \, ds, \qquad (3.23)$$

where $\phi(s)$ is an auxiliary function to be determined. Substituting the relation (3.23) into equation (3.22), interchanging the order of integration and then using the result

$$\int_0^\infty \tau^{-1} \, \tanh c\tau \, \cos \tau t \, \cos \tau y \, d\tau = \frac{1}{2} \log \left| \frac{\cosh \frac{\pi t}{2c} + \cosh \frac{\pi y}{2c}}{\cosh \frac{\pi t}{2c} - \cosh \frac{\pi y}{2c}} \right|,$$

where $c > 0$, $t > 0$ and $y > 0$, we find

$$\frac{1}{\pi} \int_0^a \frac{\cosh \frac{\pi s}{2c} \, \phi(s)}{\cosh \frac{\pi s}{c} - \cosh \frac{\pi t}{c}} \, ds = \frac{c \, g_1'(t)}{\pi^2 \, \sinh \frac{\pi t}{2c}}, \ 0 < t < a, \qquad (3.24)$$

where $g_1'(t)$ denotes the differentiation of $g_1(t)$ with respect to t, and the integral is in the sense of the Cauchy principal value. This is a first-kind integral equation with

96

Cauchy kernel. Its solution is obtained from standard results and is given by (cf. Cooke 1970)

$$\phi(s) = -\frac{4\sqrt{2}}{\pi c} \frac{\sinh^2(\frac{\pi s}{2c})}{\{\cosh(\frac{a\pi}{c}) - \cosh(\frac{\pi s}{c})\}^{1/2}}$$

$$\times \int_0^a \left\{ \frac{\cosh(\frac{a\pi}{c}) - \cosh(\frac{\pi t}{c})}{\cosh(\frac{\pi t}{c}) - 1} \right\}^{\frac{1}{2}} \frac{\cosh(\frac{\pi t}{2c}) \, g_1'(t)}{\{\cosh(\frac{\pi t}{c}) - \cosh(\frac{\pi s}{c})\}} \, dt$$

$$+ \frac{\pi \sqrt{2} \, C_1}{c \{\cosh(\frac{a\pi}{c}) - \cosh(\frac{\pi s}{c})\}^{\frac{1}{2}}}, \quad 0 < s < a, \tag{3.25}$$

where C_1 is an arbitrary constant. Substituting the value of $f(\tau)$ from the relation (3.23) into (3.21), interchanging the order of integration and then using the result (A.2.3), we find that

$$\int_0^a \frac{\phi(s) \, ds}{\sqrt{\cosh x - \cosh s}} = 0, \quad x > a. \tag{3.26}$$

Multiplying both sides of (3.26) by $(\sinh \frac{x}{2} / \cosh^2 \frac{x}{2})$ and then integrating it with respect to x between a and ∞, we get

$$\sqrt{2} \int_0^a \frac{\phi(s) \, ds}{\cosh^2(\frac{s}{2})} - \int_0^a \frac{\sqrt{\cosh a - \cosh s}}{\cosh^2(\frac{s}{2}) \, \cosh(\frac{a}{2})} \phi(s) \, ds = 0. \tag{3.27}$$

Now putting the value of $\phi(s)$ from (3.25) into (3.27), the unknown arbitrary constant C_1 is found to be given by

$$C_1 = \frac{1}{(I_1 - I_2)} \left[\frac{8}{2c} \int_0^a \frac{\sinh^2(\frac{\pi s}{2c}) ds}{\cosh^2(\frac{s}{2}) \{\cosh(\frac{a\pi}{c}) - \cosh(\frac{\pi s}{c})\}^{\frac{1}{2}}} \right.$$

$$\times \int_0^a \left\{ \frac{\cosh(\frac{a\pi}{c}) - \cosh(\frac{\pi t}{c})}{\cosh(\frac{\pi t}{c}) - 1} \right\}^{\frac{1}{2}} \frac{\cosh(\frac{\pi t}{2c}) \, g_1'(t)}{\{\cosh(\frac{\pi t}{c}) - \cosh(\frac{\pi s}{c})\}} \, dt$$

$$- \frac{4\sqrt{2} \, \operatorname{sech}(\frac{a}{2})}{\pi c} \int_0^a \frac{\sinh^2(\frac{\pi s}{2c}) \sqrt{\cosh a - \cosh s}}{\cosh^2(\frac{s}{2}) \{\cosh(\frac{a\pi}{c}) - \cosh(\frac{\pi s}{c})\}^{\frac{1}{2}}} \, ds$$

$$\left. \times \int_0^a \left\{ \frac{\cosh(\frac{a\pi}{c}) - \cosh(\frac{\pi t}{c})}{\cosh(\frac{\pi t}{c}) - 1} \right\}^{\frac{1}{2}} \frac{\cosh(\frac{\pi t}{2c}) \, g_1'(t)}{\{\cosh(\frac{\pi t}{c}) - \cosh(\frac{\pi s}{c})\}} \, dt \right], \tag{3.28}$$

where

$$I_1 = \frac{2\pi}{c} \int_0^a \frac{ds}{\cosh^2(\frac{s}{2}) \left\{ \cosh(\frac{a\pi}{c}) - \cosh(\frac{\pi s}{c}) \right\}^{\frac{1}{2}}},$$

$$I_2 = \frac{\pi \sqrt{2} \operatorname{sech} \left(\frac{a}{2}\right)}{c} \int_0^a \frac{\sqrt{\cosh a - \cosh s} \; ds}{\cosh^2(\frac{s}{2}) \left\{ \cosh(\frac{a\pi}{c}) - \cosh(\frac{\pi s}{c}) \right\}^{\frac{1}{2}}}. \tag{3.29}$$

Thus, the explicit solution of the dual integral equations (3.20) and (3.21) is obtained by the relations (3.23), (3.25) and (3.28).

For some particular value of c, the integrals in (3.29) can be calculated. As an example, we consider the particular case of $g(x) = 1$ to find the values of $f(\tau)$ for which $c = \pi$. Then, we have

$$g_1'(t) = \frac{1}{2} \sinh\left(\frac{t}{2}\right).$$

From (3.25), we find that

$$\phi(s) = -\frac{\sqrt{2}}{\pi} \frac{\sinh^2(\frac{s}{2})}{\sqrt{\cosh a - \cosh s}} + \frac{\sqrt{2} \, C_1}{\sqrt{\cosh a - \cosh s}}, \quad 0 < s < a, \tag{3.30}$$

where the constant C_1 is obtained from (3.28):

$$C_1 = \left\{ \tanh\left(\frac{a}{2}\right) + E\left(\frac{\pi}{2}, \tanh\left(\frac{\pi}{2}\right)\right) \right\}^{-1}$$

$$\times \left[\frac{1}{\pi}\left(a - 2 \tanh\frac{a}{2}\right) - \frac{1}{\pi} E\left(\frac{\pi}{2}, \tanh\frac{a}{2}\right) - F\left(\frac{\pi}{2}, \tanh\frac{a}{2}\right) \right], \tag{3.31}$$

with

$$I_1 = -2\sqrt{2} \operatorname{sech}\left(\frac{a}{2}\right) E\left(\frac{\pi}{2}, \tanh\left(\frac{a}{2}\right)\right),$$

$$I_2 = 2\sqrt{2} \operatorname{sech}\left(\frac{a}{2}\right) \tanh\left(\frac{a}{2}\right),$$

where F and E are the elliptic integrals of the first and second kind respectively.

Thus the value of $f(\tau)$ can be found from the relation (3.23) where $\phi(s)$ is given by (3.30) and (3.31).

3.1.2 So far we have considered solutions of two particular pairs of integral equations in explicit form. Now we apply a special method presented in §1.2.2 to find the solutions of three new pairs of integral equations (cf. Mandal and Mandal 1993). Except for one pair, these dual integral equations are first reduced to solving some appropriate ODEs. In the solutions of these ordinary differential equations, we invoke the inversion formulae for Abel integral equations to obtain explicit solutions to these dual integral equations in two cases and in the other case they are reduced to an appropriate Fredholm integral equation of the second kind. As an example of the application of these dual integral equations, a problem arising from mathematical physics is also given.

(i) The first pair is

$$\int_0^\infty \tau^2 \, f(\tau) \, P_{-\frac{1}{2}+i\tau}(\cosh x) \, d\tau \;=\; g(x), \; 0 \le x \le a, \qquad (3.32)$$

$$\int_0^\infty \tau \, \tanh \pi\tau \, f(\tau) \, P_{-\frac{1}{2}+i\tau}(\cosh x) \, d\tau \;=\; 0, \; a \le x < \infty. \qquad (3.33)$$

To solve this pair, we assume that

$$\int_0^\infty \tau \, \tanh \pi\tau \, f(\tau) \, P_{-\frac{1}{2}+i\tau}(\cosh x) \, d\tau = \phi(x), \; 0 \le x \le a, \qquad (3.34)$$

where $\phi(x)$ is an unknown function for $0 \le x \le a$. It follows from (3.33) and (3.34), by using Mehler−Fock transform theorem (cf. Mandal and Mandal 1997, p. 21), that

$$f(\tau) = \int_0^a \phi(x) \, P_{-\frac{1}{2}+i\tau}(\cosh x) \, \sinh x \, dx. \qquad (3.35)$$

Substituting this expression for $f(\tau)$ into (3.32), we find that

$$\int_0^\infty \tau^2 \, P_{-\frac{1}{2}+i\tau}(\cosh x) \left\{ \int_0^a \phi(t) \, P_{-\frac{1}{2}+i\tau}(\cosh x) \, \sinh x \, dx \right\} d\tau = g(x), \; 0 \le x \le a. \qquad (3.36)$$

Equation (3.36) is now equivalent to the ODE

$$\frac{1}{\sinh x} \frac{d}{dx} \left(\sinh x \, \frac{du}{dx} \right) + \frac{1}{4} u = -g(x), \qquad (3.37)$$

where

$$u(x) = \int_0^\infty P_{-\frac{1}{2}+i\tau}(\cosh x) \left\{ \int_0^a \phi(t) \, P_{-\frac{1}{2}+i\tau} \, (\cosh t) \, \sinh t \, dt \right\} d\tau, \; 0 \le x \le a. \qquad (3.38)$$

The solution of the ODE (3.37), by the method of variation of parameters is

$$
\begin{aligned}
u(x) \;=\; & C_1\, P_{-\frac{1}{2}}(\cosh x) + C_2\, Q_{-\frac{1}{2}}(\cosh x) \\
& + P_{-\frac{1}{2}}(\cosh x) \int_0^x g(t)\, Q_{-\frac{1}{2}}(\cosh t)\, \sinh t\; dt \\
& - Q_{-\frac{1}{2}}(\cosh x) \int_0^x g(t)\, P_{-\frac{1}{2}}(\cosh t)\, \sinh t\; dt,
\end{aligned}
\tag{3.39}
$$

where the constants C_1 and C_2 are arbitrary. They are to be determined from physical considerations of the problem. Of these, the constant C_2 must be zero in order that $u(x)$ is finite at $x = 0$. The constant C_1 will be determined later.

Interchanging the order of integration and then using the result (A.2.4), and again interchanging the order of integration, equation (3.38) reduces to an Abel integral equation

$$
\int_0^x \frac{1}{\sqrt{\cosh x - \cosh s}} \left\{ \int_s^a \frac{\sinh t\; \phi(t)}{\sqrt{\cosh t - \cosh s}}\; dt \right\}\; ds = \pi\, u(x),\quad 0 \le x \le a,
$$

so that

$$
\int_x^a \frac{\sinh t\; \phi(t)}{\sqrt{\cosh t - \cosh x}}\; dt = \psi(x),
$$

where

$$
\psi(x) = \sinh x \int_0^x \frac{u'(s)\; ds}{\sqrt{\cosh x - \cosh s}}.
\tag{3.40}
$$

Another use of Abel's inversion formula gives

$$
\sinh x\; \phi(x) = -\frac{1}{\pi} \frac{d}{dx} \int_x^a \frac{\sinh s\; \psi(s)}{\sqrt{\cosh s - \cosh x}}\; ds.
\tag{3.41}
$$

This ultimately produces the complete solution of the pair (3.32) and (3.33) provided the constant C_1 is determined. This is found from the fact that $\psi(a) = 0$, which arises from the physical requirement involving the continuity of $\phi(x)$ at $x = a$. Hence, the constant C_1 is determined by the relation

$$
\int_0^a \frac{u'(s)\; ds}{\sqrt{\cosh a - \cosh s}} = 0.
\tag{3.42}
$$

(ii) The second pair is

$$
\int_0^\infty (\tau^2 - \mu^2)\, f(\tau)\, P_{-\frac{1}{2}+i\tau}(\cosh x)\, d\tau = g(x),\quad 0 \le x \le a,
\tag{3.43}
$$

$$
\int_0^\infty \tau \tanh \pi\tau\, f(\tau)\, P_{-\frac{1}{2}+i\tau}(\cosh x)\, d\tau = 0,\quad a \le x < \infty,
\tag{3.44}
$$

100

where μ is a real constant. This pair is a generalization of the pair (3.32) and (3.33) in the sense that for $\mu = 0$, this is reduced to the pair (3.32) and (3.33).

To solve this pair, we assume that

$$\int_0^\infty \tau \tanh \pi\tau \; f(\tau) \; P_{-\frac{1}{2}+i\tau}(\cosh x) \; d\tau = \phi(x), \; 0 \le x \le a, \qquad (3.45)$$

where $\phi(x)$ is unknown for $0 \le x \le a$. Thus

$$f(\tau) = \int_0^a \phi(t) \; P_{-\frac{1}{2}+i\tau}(\cosh t) \; \sinh t \; dt. \qquad (3.46)$$

Substituting (3.46) into (3.43), we find that

$$\int_0^\infty (\tau^2 - \mu^2) \; P_{-\frac{1}{2}+i\tau}(\cosh x) \left\{ \int_0^a \phi(t) \; P_{-\frac{1}{2}+i\tau}(\cosh t) \sinh t \; dt \right\} d\tau = g(x), \qquad (3.47)$$

for $0 \le x \le a$. This equation is equivalent to the ODE

$$\frac{1}{\sinh x} \frac{d}{dx} \left(\sinh x \; \frac{du}{dx} \right) + \left(\frac{1}{4} + \mu^2 \right) u = -g(x), \; 0 \le x \le a, \qquad (3.48)$$

where now

$$u(x) = \int_0^\infty P_{-\frac{1}{2}+i\tau} (\cosh x) \left\{ \int_0^a \phi(t) \; P_{-\frac{1}{2}+i\tau} (\cosh t) \sinh t \; dt \right\} d\tau. \qquad (3.49)$$

The solution of the ODE is

$$\begin{aligned}
u(x) \;=\;& C_1 \; P_{-\frac{1}{2}+i\mu}(\cosh x) + C_2 \; Q_{-\frac{1}{2}+i\mu}(\cosh x) \\
& + P_{-\frac{1}{2}+i\mu}(\cosh x) \int_0^x g(s) \; Q_{-\frac{1}{2}+i\mu}(\cosh s) \sinh s \; ds \\
& - Q_{-\frac{1}{2}+i\mu}(\cosh x) \int_0^x g(s) \; P_{-\frac{1}{2}+i\mu}(\cosh s) \sinh s \; ds.
\end{aligned} \qquad (3.50)$$

Since $u(x)$ must be finite at $x = 0$, the constant C_2 must be equal to zero. Now, proceeding as above, the solution of this pair can be obtained easily.

(iii) Lastly, we consider the dual integral equations

$$\int_0^\infty \coth \pi\tau \; f(\tau) \; P_{-\frac{1}{2}+i\tau}(\cosh x) \; d\tau \;=\; g(x), \; 0 \le x \le a, \quad (3.51)$$

$$\int_0^\infty \left[\lambda\tau + \mu(\tau^2 + c^2) \coth \pi\tau \right] \; f(\tau) \; P_{-\frac{1}{2}+i\tau}(\cosh x) \; d\tau \;=\; 0, \; a \le x < \infty, \quad (3.52)$$

101

where λ, μ, c are real constants and $c > \frac{1}{2}$.

.When $\mu = 0$, this pair reduces to the pair considered by Babloian (1964) (writing $F(\tau) = \coth \pi\tau \ f(\tau)$).

When $\lambda = 0$ but $\mu \neq 0$, this pair becomes

$$\int_0^\infty \coth \pi\tau \ f(\tau) \ P_{-\frac{1}{2}+i\tau}(\cosh x) \ d\tau \ = \ g(x), \ 0 \leq x \leq a, \quad (3.53)$$

$$\int_0^\infty (\tau^2 + c^2) \ \coth \pi\tau \ f(\tau) \ P_{-\frac{1}{2}+i\tau}(\cosh x) \ d\tau \ = \ 0, \ a \leq x < \infty, \quad (3.54)$$

whose solution can be obtained in a similar manner as in case (i), and the solution in this case is given by (after utilizing the result (2) on p. 170 of Erdélyi et al. (1953) involving Legendre functions in the simplification)

$$\begin{aligned}
f(\tau) \ = \ & \tau \ \tanh^2 \pi\tau \int_0^a g(x) \ P_{-\frac{1}{2}+i\tau}(\cosh x) \ \sinh x \ dx \\
& + \frac{\tau \ \tanh^2 \pi\tau \ \sinh a \ f(x)}{(\tau^2 + c^2) \ Q_{c-\frac{1}{2}}(\cosh a)} \ \left[Q_{c-\frac{1}{2}}(\cosh a) \ P^1_{-\frac{1}{2}+i\tau}(\cosh a) \right. \\
& \left. - P_{-\frac{1}{2}+i\tau}(\cosh a) \ Q^1_{c-\frac{1}{2}}(\cosh a) \right].
\end{aligned} \quad (3.55)$$

When both λ and μ are non-zero, we can write (3.52) as

$$\int_0^\infty (\tau^2 + c^2) \ \left[1 + \frac{\lambda \ \tau}{\mu(\tau^2 + c^2)} \ \tanh \pi\tau \right] \coth \pi\tau$$

$$\times \ f(\tau) \ P_{-\frac{1}{2}+i\tau}(\cosh x) \ d\tau = 0, \ a \leq x < \infty. \quad (3.56)$$

This is equivalent to the ODE

$$\frac{1}{\sinh x} \ \frac{d}{dx} \ \left(\sinh x \ \frac{du}{dx} \right) + \left(\frac{1}{4} - c^2 \right) u = 0, \ a \leq x < \infty, \quad (3.57)$$

where

$$u(x) = \int_0^\infty \left[1 + \frac{\lambda \ \tau}{\mu(\tau^2 + c^2)} \ \tanh \pi\tau \right] \coth \pi\tau \ f(\tau) \ P_{-\frac{1}{2}+i\tau}(\cosh x) \ d\tau. \quad (3.58)$$

The solution of the ODE (3.57) is

$$u(x) = C_1 \ P_{c-\frac{1}{2}} (\cosh x) + C_2 \ Q_{c-\frac{1}{2}}(\cosh x), \ a \leq x < \infty.$$

102

Since $u(x)$ must remain finite as $x \to \infty$ and $c > \frac{1}{2}$, we must have $C_1 = 0$. Thus equation (3.58) produces

$$\int_0^\infty f(\tau) \, \coth \pi\tau \, P_{-\frac{1}{2}+i\tau} (\cosh x) \, d\tau$$

$$= C_2 \, Q_{c-\frac{1}{2}}(\cosh x) - \frac{\lambda}{\mu} \int_0^\infty \frac{\gamma \, f(\gamma)}{(\gamma^2 + c^2)} \, P_{-\frac{1}{2}+i\gamma}(\cosh x) \, d\gamma, \quad a \le x < \infty. \qquad (3.59)$$

Equations (3.51) and (3.59) give $f(\tau)$ by the Mehler−Fock inversion theorem. After some simplification, we find that

$$\frac{f(\tau)}{\tau} = C_1 \frac{\tanh^2 \pi\tau \, \sinh a}{(\tau^2 + c^2)} \left[Q_{c-\frac{1}{2}}(\cosh x) \, P^1_{-\frac{1}{2}+i\tau}(\cosh a) - P_{-\frac{1}{2}+i\tau}(\cosh a) \right.$$

$$\left. \times Q^1_{c-\frac{1}{2}}(\cosh a) \right] + \tanh^2 \pi\tau \int_0^a g(x) \, P_{-\frac{1}{2}+i\tau} (\cosh x) \, \sinh x \, dx$$

$$- \frac{\lambda \, \tanh^2 \pi\tau}{\mu} \left[\frac{f(\tau)}{(\tau^2 + c^2) \tanh \pi\tau} - \int_0^\infty \frac{\gamma \, f(\gamma) \, \sinh a}{(\gamma^2 + c^2) \, (\tau^2 - \gamma^2)} \right.$$

$$\times \left\{ P_{-\frac{1}{2}+i\tau}(\cosh a) \, P^1_{-\frac{1}{2}+i\gamma}(\cosh a) - P_{-\frac{1}{2}+i\gamma}(\cosh a) \right.$$

$$\left. \left. \times P^1_{-\frac{1}{2}+i\tau}(\cosh a) \right\} d\gamma \right]. \qquad (3.60)$$

The continuity requirement at $x = a$, gives

$$C_2 = \frac{g(a)}{Q_{c-\frac{1}{2}}(\cosh a)} + \frac{\lambda}{\mu \, Q_{c-\frac{1}{2}}(\cosh a)} \int_0^\infty \frac{\gamma \, f(\gamma)}{(\gamma^2 + c^2)} \, P_{-\frac{1}{2}+i\gamma}(\cosh a) \, d\gamma. \qquad (3.61)$$

Using the above relation in (3.60), we obtain the Fredholm integral equation of the second kind for $f(\tau)$ as

$$\frac{f(\tau)}{\tau} - \lambda \int_0^\infty \frac{f(\gamma)}{\gamma} \, K(\gamma, \tau) \, d\gamma = L(\tau), \qquad (3.62)$$

where

$$K(\gamma, \tau) = \frac{\gamma^2 \, \sinh a \, \tanh^2 \pi\tau}{(\gamma^2 + c^2) \{\lambda\tau \, \tanh \pi\tau + \mu \, (\tau^2 + c^2)\}}$$

$$\times \left[\frac{P_{-\frac{1}{2}+i\gamma}(\cosh a)}{Q_{c-\frac{1}{2}}(\cosh a)} \left\{ P^1_{-\frac{1}{2}+i\tau}(\cosh a) \, Q_{c-\frac{1}{2}}(\cosh a) \right. \right.$$

$$\left. - P_{-\frac{1}{2}+i\tau}(\cosh a) \, Q^1_{c-\frac{1}{2}}(\cosh a) \right\}$$

$$+ (\tau^2 + c^2)(\tau^2 - \gamma^2)^{-1} \left\{ P^1_{-\frac{1}{2}+i\gamma}(\cosh a) \, P_{-\frac{1}{2}+i\tau}(\cosh a) \right.$$

$$\left. \left. - P_{-\frac{1}{2}+i\gamma}(\cosh a) \, P^1_{-\frac{1}{2}+i\tau}(\cosh a) \right\} \right],$$

103

and

$$L(\tau) = \frac{\mu \, \tanh^2 \pi\tau}{\{\lambda \, \tau \, \tanh \pi\tau + \mu(\tau^2 + c^2)\}} \left[\frac{\sinh a}{Q_{c-\frac{1}{2}}(\cosh a)} \right.$$
$$\times \left\{ P^1_{-\frac{1}{2}+i\tau}(\cosh a) Q_{c-\frac{1}{2}}(\cosh a) - P_{-\frac{1}{2}+i\tau}(\cosh a) Q^1_{c-\frac{1}{2}}(\cosh a) \right\} g(a)$$
$$\left. + (\tau^2 + c^2) \int_0^a g(x) \, P_{-\frac{1}{2}+i\tau}(\cosh x) \, \sinh x \, dx \right].$$

As applications of these dual integral equations, we consider a mixed boundary value problem involving the displacement in a half-space due to torsion of an attached rigid annular die. Using toroidal coordinates (α, β, θ) where

$$r = \sqrt{x^2 + y^2} = \frac{c \, \sinh \alpha}{\cosh \alpha + \cos \beta}, \quad z = \frac{c \, \sin \beta}{\cosh \alpha + \cos \beta} \quad (c > 0),$$

$0 \le \alpha < \infty$, $0 \le \beta \le \pi$, the half-space is $z \ge 0$, and the die is represented by $z = 0$, $\alpha_0 < \alpha < \infty$. The state of stress and strain does not depend on the angular coordinate θ and is determined by the non-zero component of displacement $u_0 = u(r, z)$ and satisfies

$$\frac{\partial^2 u}{\partial r^2} + \frac{1}{r} \frac{\partial u}{\partial r} + \frac{\partial^2 u}{\partial z^2} - \frac{u}{r^2} = 0, \ z \ge 0$$

with the boundary conditions

$$\frac{\partial u}{\partial \beta} \big|_{\beta=0} = 0, \ 0 \le \alpha \le \alpha_0, \ u \big|_{\beta=0} = kr, \ \alpha_0 < \alpha < \infty, \ u \big|_{\beta=\pi} = 0,$$

where k is the angle of rotation of the die.

We seek a solution of this mixed boundary value problem in the form

$$u = k \, c \, (\cosh \alpha + \cos \beta)^{\frac{1}{2}} \int_0^\infty f(\tau) \frac{\sinh(\pi - \beta)\tau}{\cosh \pi\tau} \, P^1_{-\frac{1}{2}+i\tau}(\cosh \alpha) \, d\tau;$$

then the boundary conditions produce the dual integral equations

$$\int_0^\infty \tau \, f(\tau) \, P^1_{-\frac{1}{2}+i\tau}(\cosh \alpha) \, d\tau = 0, \ 0 \le \alpha \le \alpha_0,$$

$$\int_0^\infty \tanh \pi\tau \, f(\tau) \, P^1_{-\frac{1}{2}+i\tau}(\cosh \alpha) \, d\tau = \frac{\tanh \alpha/2}{\sqrt{\cosh \alpha + 1}}, \ \alpha_0 < \alpha < \infty,$$

for the unknown function $f(\tau)$. Using the formula $P^1_{-\frac{1}{2}+i\tau}(\cosh \alpha) = \frac{d}{d\alpha} P_{-\frac{1}{2}+i\tau}$ $(\cosh \alpha)$ and taking $f(\tau) = \tau \, A(\tau)$, we obtain the dual integral equations discussed in case (i) and hence the solution is obvious.

104

3.2 Kernels involving associated Legendre functions

In this section, we consider some dual integral equations with associated Legendre function kernel. Rukhovets and Ufliand (1966) considered such a pair of integral equations involving $P^m_{-\frac{1}{2}+i\tau}(\cosh x)$ $(m = 0, 1, 2, \ldots)$ together with their applications to some physical problems in the theory of elasticity.

3.2.1 We consider the dual integral equations

$$\int_0^\infty [1 + w(\tau)] \, f(\tau) \, P^m_{-\frac{1}{2}+i\tau}(\cosh x) \, d\tau \;=\; g(x), \; 0 \le x < a, \qquad (3.63)$$

$$\int_0^\infty \tau \, \tanh \pi\tau \, f(\tau) \, P^m_{-\frac{1}{2}+i\tau}(\cosh x) \, d\tau \;=\; 0, \; a < x < \infty, \qquad (3.64)$$

where $w(\tau)$ is a known weight function. To solve the above pair, let us assume that

$$f(\tau) = \int_0^a \phi(t) \, \cos \tau t \, dt, \qquad (3.65)$$

where $\phi(t)$ is an unknown function which has continuous derivative in $(0, a)$.

Using the relations

$$P^m_{-\frac{1}{2}+i\tau}(\lambda) = (\lambda^2 - 1)^{\frac{m}{2}} \frac{d^m}{d\lambda^m} \left\{ P_{-\frac{1}{2}+i\tau}(\lambda) \right\}, \; \lambda = \cosh x, \qquad (3.66)$$

and

$$\int_0^\infty \tanh \pi\tau \, P_{-\frac{1}{2}+i\tau}(\cosh x) \sin \tau t \, d\tau = \begin{cases} 0, & t < x, \\ [2(\cosh t - \cosh x)]^{-\frac{1}{2}}, & t > x, \end{cases} \qquad (3.67)$$

it is observed that equation (3.64) is satisfied identically. Integrating both sides of (3.63) m times with respect to λ ($= \cosh x$), it reduces to the form

$$\int_0^\infty [1 + w(\tau)] \, f(\tau) \, P_{-\frac{1}{2}+i\tau}(\cosh x) \, d\tau = G(x) + \sum_{k=0}^{m-1} C_k \, \lambda^k = \chi(x), \; \text{say}, \qquad (3.68)$$

where the C_k's $(k = 0, 1, \ldots, m - 1)$ are arbitrary constants and

$$G(x) = \int_1^\lambda \int_1^\lambda \cdots \int_1^\lambda (\lambda^2 - 1)^{-\frac{m}{2}} g(x) \, d\lambda^m. \qquad (3.69)$$

105

Substituting (3.65) into (3.68) and then using the result (A.2.3) we find an Abel integral equation

$$\int_0^x \frac{\Phi(t)}{[2(\cosh x - \cosh t)]^{\frac{1}{2}}} \, dt = \chi(x), \tag{3.70}$$

where

$$\phi(t) + \frac{1}{\pi} \int_0^a [\psi(s+t) + \psi(s-t)] \, \phi(s) \, ds = \Phi(t), \tag{3.71}$$

with

$$\psi(y) = \int_0^\infty w(\tau) \, \cos \tau y \, d\tau. \tag{3.72}$$

Since the solution of the integral equation (3.70) is given by

$$\Phi(t) = \frac{2}{\pi} \frac{d}{dt} \int_0^t [2(\cosh t - \cosh x)]^{-\frac{1}{2}} \, \chi(x) \, \sinh x \, dx, \tag{3.73}$$

the solution of the dual integral equations is reduced to the solution of equation (3.71) which is a Fredholm integral equation of the second kind. The solution of this integral equation contains m arbitrary constants which can be found by imposing supplementary conditions. For this, we assume that

$$h(x) = \int_0^\infty \tau \, f(\tau) \, \tanh \pi\tau \, P_{-\frac{1}{2}+i\tau}^m(\cosh x) \, d\tau, \; 0 \le x < a. \tag{3.74}$$

Substituting (3.65) into (3.74) and then using the relations (3.66) and (3.67), we get

$$\begin{aligned}
h(x) &= (\lambda^2 - 1)^{\frac{m}{2}} \frac{d^m}{d\lambda^m} \Big\{ \phi(a)[2(\cosh a - \cosh x)]^{-\frac{1}{2}} \\
&\quad - \int_x^a \phi'(t) \, [2(\cosh t - \cosh x)]^{-\frac{1}{2}} \, dt \Big\}, \; 0 \le x < a.
\end{aligned} \tag{3.75}$$

After successive integration by parts, (3.75) is equivalent to the condition that

$$\phi^{(i)}(a) = 0; \; i = 0, 1, 2, \ldots, m-1, \tag{3.76}$$

which follows from the fact that as $x \to a - 0$, the function $h(x)$ assumes the order of $(\cosh a - \cosh x)^{-\frac{1}{2}}$. It can be easily seen that the condition (3.76) determines the constants C_k. We represent $\phi(x)$ in the form

$$\phi(x) = \phi_1(x) + \phi_2(x), \; \phi_2(x) = \sum_{k=0}^{m-1} C_k \, p_k(x), \tag{3.77}$$

106

where $\phi_1(x)$ and $p_k(x)$ $(k = 0, 1, 2, \ldots, m-1)$ are unknown functions.

Set

$$
\begin{aligned}
\Phi(x) &= \Phi_1(x) + \Phi_2(x), \quad \text{where} \\
\Phi_1(x) &= \frac{2}{\pi} \frac{d}{dx} \int_0^x \frac{\sinh t \, G(t)}{[2(\cosh x - \cosh t)]^{\frac{1}{2}}} \, dt, \quad (3.78) \\
\Phi_2(x) &= \sum_{k=0}^{m-1} C_k \frac{dI_k}{dx}, \quad (3.79)
\end{aligned}
$$

with

$$
I_k(x) = \frac{2}{\pi} \int_0^x [2(\cosh x - \cosh t)]^{-\frac{1}{2}} \cosh^k t \, \sinh t \, dt.
$$

Then from the integral equation (3.71), we get different Fredholm integral equations with given right hand sides $\Phi_1(x)$ and $\dfrac{dI_k}{dx}$ for the unknown functions $\phi_1(x)$ and $p_k(x)$. Thus the conditions (3.76) now become the following linear algebraic system of equations for the unknown constants C_k:

$$
\phi_1^{(i)}(a) + \sum_{k=0}^{m-1} C_k \, p_k^{(i)}(a) = 0; \quad i = 0, 1, 2, \ldots, m-1. \quad (3.80)
$$

As an example of an application of the dual integral equations presented above, we consider a boundary value problem (cf. Rukhovets and Ufliand 1966) for a spherical segment when the required harmonic function $u(r, \theta, z)$ vanishes on the spherical surface and satisfies the mixed conditions

$$
u = u_1(r, \theta), \ 0 \leq r < a; \ \frac{\partial u}{\partial z} = 0, \ a < r < b, \quad (3.81)
$$

on the flat surface $z = 0$ $(r, \theta, z$ being the cylindrical coordinate system).

We consider the toroidal coordinate system

$$
x = \frac{b \, \sinh \alpha \, \cos \theta}{\cosh \alpha + \cos \beta}, \ y = \frac{b \, \sinh \alpha \, \sin \theta}{\cosh \alpha + \cos \beta}, \ z = \frac{b \, \sin \beta}{\cosh \alpha + \cos \beta},
$$

where $0 \leq \alpha < \infty$, $0 \leq \beta \leq \gamma$. Then the equation of the spherical surface of the segment will be $\beta = \gamma$. On the flat surface $\beta = 0$, the line of division between boundary conditions of the first and second kind is defined by the circle $\alpha = \alpha_0$ $(\tanh \frac{\alpha_0}{2} = \frac{b}{a})$, and at $z = 0$, $r \to b$ we have $\alpha \to \infty$.

107

We now seek a solution to the problem which satisfies the condition $u\mid_{\beta=\gamma}= 0$ in the form

$$u = \sqrt{\cosh\alpha + \cos\beta} \sum_{m=-\infty}^{\infty} e^{im\theta} \int_0^\infty B(\tau) \frac{\sinh(\gamma - \beta)\tau}{\sinh\gamma\tau} P_{-\frac{1}{2}+i\tau}^m(\cosh\alpha)\, d\tau. \quad (3.82)$$

Expressing the prescribed function $u_1(r,\theta)$ in a Fourier series in the angular coordinate

$$u_1(r,\theta) = \sqrt{\cosh\alpha + 1} \sum_{m=-\infty}^{\infty} g(\alpha)\, e^{im\theta},$$

the above problem reduces to the solution of the following dual integral equations

$$\int_0^\infty B(\tau)\, P_{-\frac{1}{2}+i\tau}^m(\cosh\alpha)\, d\tau = g(x),\ 0 \leq \alpha < \alpha_0, \quad (3.83)$$

$$\int_0^\infty \tau\, \coth\gamma\tau\, B(\tau)\, P_{-\frac{1}{2}+i\tau}^m(\cosh\alpha)\, d\tau = 0,\ \alpha_0 < \alpha < \infty. \quad (3.84)$$

By the substitution $B(\tau)\coth\gamma\tau = f(\tau)\tanh\pi\tau$, the above pair reduces to the pair (3.63) and (3.64) with the weight function

$$\omega(t) = -\frac{\cosh(\pi - \gamma)\tau}{\cosh\pi\tau\ \cosh\gamma\tau}.$$

When $m = 1$, this pair arises in the problem of the torsion of a truncated sphere (cf. Babloian 1964).

3.2.2 The dual integral equations discussed above and in the previous section can be generalized to the case when the kernels involve the associated Legendre function $P_{-\frac{1}{2}+i\tau}^\mu(\cosh x)$ where μ is a complex number such that $|Re\ \mu| < \frac{1}{2}$. Pathak (1978) studied the following dual integral equations involving associated Legendre functions of the first kind with complex superscript. These are also mentioned in the book of Virchenko (1989) p. 85, as a special case of some triple integral equations.

(i) First we consider the dual integral equations

$$\int_0^\infty f(\tau)\, P_{-\frac{1}{2}+i\tau}^\mu(\cosh x)\, d\tau = g(x),\ 0 \leq x \leq a, \quad (3.85)$$

$$\int_0^\infty f(\tau)\, \Gamma\left(\frac{1}{2} - \mu + i\tau\right)\Gamma\left(\frac{1}{2} - \mu - i\tau\right)\tau\,\sinh\pi\tau$$

$$\times P_{-\frac{1}{2}+i\tau}^\mu(\cosh x)\, d\tau = h(x),\ x > a. \quad (3.86)$$

Multiplying (3.85) by $\pi^{-1}\Gamma(\frac{1}{2} - \mu) \cos \pi\mu \ \sinh^{1-\mu} x(\cosh t - \cosh x)^{\mu-\frac{1}{2}}$, integrating the relation with respect to x from 0 to t, and then differentiating with respect to t, we obtain

$$\pi^{-1}\ \Gamma\left(\frac{1}{2} - \mu\right)\cos\pi\mu \int_0^\infty\ f(\tau)\left\{\frac{d}{dt}\int_0^t \frac{\sinh^{1-\mu} x\ P_{-\frac{1}{2}+i\tau}^\mu(\cosh x)}{(\cosh t - \cosh x)^{\frac{1}{2}-\mu}}\ dx\right\}$$

$$\times\ P_{-\frac{1}{2}+i\tau}^\mu(\cosh x)\ d\tau = G_1(t),\ 0 \le t \le a, \tag{3.87}$$

where

$$G_1(t) = \pi^{-1}\ \Gamma\left(\frac{1}{2} - \mu\right)\ \cos\pi\mu\ \frac{d}{dt}\int_0^t \frac{\sinh^{1-\mu} x\ g(x)}{(\cosh t - \cosh x)^{\frac{1}{2}-\mu}}\ dx. \tag{3.88}$$

Using the result (A.2.18), the equation (3.87) becomes

$$F_c[f(\tau)] = G_1(t),\ 0 \le t \le a, \tag{3.89}$$

where F_c denotes the Fourier cosine transform.

Again, we multiply (3.86) by $\pi^{-2}\ \Gamma(\frac{1}{2} + \mu)\cos\pi\mu\ \sinh^{1+\mu} x\ (\cosh x - \cosh t)^{-\frac{1}{2}-\mu}$, integrate with respect to x from t to ∞ and then apply the result (A.2.21), to reduce it to the form

$$F_c[f(\tau)] = G_2(t),\ t > a, \tag{3.90}$$

where

$$G_2(t) = \pi^{-2}\ \Gamma\left(\frac{1}{2} + \mu\right)\ \cos\pi\mu\ \int_t^\infty h(x)\ \sinh^{1+\mu} x(\cosh x - \cosh t)^{-\mu-\frac{1}{2}}\ dx. \tag{3.91}$$

Therefore, by the Fourier cosine inversion theorem, we get

$$f(\tau) = \sqrt{\frac{2}{\pi}}\left\{\int_0^1 G_1(t)\ \cos\tau t\ dt + \int_a^\infty G_2(t)\ \cos\tau t\ dt\right\}. \tag{3.92}$$

In many physical problems, the function $g(x)$ for $x > a$ and the function $h(x)$ for $0 \le x \le a$ are needed. From (3.85), we find that for $x > a$,

109

$$g(x) = \int_0^\infty P^\mu_{-\frac{1}{2}+i\tau}(\cosh x) \left\{ \sqrt{\frac{2}{\pi}} \int_0^a G_1(t) \cos \tau t \, dt \right\} d\tau$$

$$+ P^\mu_{-\frac{1}{2}+i\tau}(\cosh x) \left\{ \sqrt{\frac{2}{\pi}} \int_a^\infty G_2(t) \cos \tau t \, dt \right\} d\tau$$

$$= \int_0^a G_1(t) \left\{ \sqrt{\frac{2}{\pi}} \int_0^\infty P^\mu_{-\frac{1}{2}+i\tau}(\cosh x) \cos \tau t \, d\tau \right\} dt$$

$$+ \int_a^\infty G_2(t) \left\{ \sqrt{\frac{2}{\pi}} \int_0^\infty P^\mu_{-\frac{1}{2}+i\tau}(\cosh x) \cos \tau t \, d\tau \right\} dt.$$

Utilizing the result (A.2.15), the above relation becomes

$$g(x) = \frac{\sinh^\mu x}{\Gamma(\frac{1}{2}-\mu)} \left[\int_0^a \frac{G_1(t) \, dt}{(\cosh x - \cosh t)^{\frac{1}{2}+\mu}} + \int_a^x \frac{G_2(t) \, dt}{(\cosh x - \cosh t)^{\frac{1}{2}+\mu}} \right], \quad x > a.$$
$$(3.93)$$

In a similar manner, from (3.86) it can be shown that for $0 \le x \le a$,

$$h(x) = \frac{\pi \sinh^{-\mu} x}{\Gamma(\frac{1}{2}+\mu)} \left[\frac{G_1(a)}{(\cosh a - \cosh x)^{\frac{1}{2}-\mu}} - \int_0^a \frac{G_1'(t) \, dt}{(\cosh t - \cosh x)^{\frac{1}{2}-\mu}} \right.$$

$$\left. - \frac{G_2(a)}{(\cosh a - \cosh x)^{\frac{1}{2}-\mu}} - \int_a^\infty \frac{G_2'(t) \, dt}{(\cosh t - \cosh x)^{\frac{1}{2}-\mu}}. \right.$$
$$(3.94)$$

(ii) Next, we consider the pair

$$\int_0^\infty \tau f(\tau) P^\mu_{-\frac{1}{2}+i\tau}(\cosh x) \, d\tau = g(x), \quad 0 \le x \le a, \qquad (3.95)$$

$$\int_0^\infty f(\tau) \Gamma\left(\frac{1}{2} - \mu + i\tau\right) \Gamma\left(\frac{1}{2} - \mu - i\tau\right) \sinh \pi \tau$$

$$\times P^\mu_{-\frac{1}{2}+i\tau}(\cosh x) \, d\tau = h(x), \quad a < x < \infty. \qquad (3.96)$$

In the process of obtaining the solutions of this pair, first we multiply (3.95) by

$$\pi^{-1}\Gamma\left(\frac{1}{2} - \mu\right) \cos \pi\mu \, \sinh^{1-\mu} x \, (\cosh t - \cosh x)^{-\frac{1}{2}+\mu}$$

and integrate with respect to x from 0 to t to obtain

$$\pi^{-1} \Gamma \left(\frac{1}{2} - \mu \right) \cos \pi \mu \int_0^\infty \tau f(\tau) \left\{ \int_0^t P_{-\frac{1}{2}+i\tau}^\mu (\cosh x) \sinh^{1-\mu} x \right.$$

$$\left. \times (\cosh t - \cosh x)^{-\frac{1}{2}+\mu} dx \right\} d\tau = G_1(t), \ 0 \le t \le a, \tag{3.97}$$

where

$$G_1(t) = \pi^{-1} \Gamma \left(\frac{1}{2} - \mu \right) \cos \pi \mu \int_0^t g(x) \sinh^{1-\mu} x (\cosh t - \cosh x)^{-\frac{1}{2}+\mu} dx. \tag{3.98}$$

Applying the result (A.2.19), equation (3.97) can be expressed as

$$F_s \left[f(\tau) \right] = G_1(t), \ 0 \le t \le a, \tag{3.99}$$

where F_s denotes the Fourier sine transform.

Now we multiply equation (3.96) by

$$\pi^{-2} \Gamma \left(\frac{1}{2} + \mu \right) \cos \pi \mu \ \sinh^{1+\mu} x \ (\cosh x - \cosh t)^{-\frac{1}{2}-\mu},$$

integrate with respect to x from t to ∞, and then differentiate with respect to t and use the relation (A.2.20). Finally we obtain

$$F_s[f(\tau)] = G_2(t), \ a < t < \infty, \tag{3.100}$$

where

$$G_2(t) = -\pi^{-2} \Gamma \left(\frac{1}{2} + \mu \right) \cos \pi \mu \ \frac{d}{dt} \int_t^\infty h(x) \sinh^{1+\mu} x \ (\cosh x - \cosh t)^{-\frac{1}{2}-\mu} dx. \tag{3.101}$$

Hence, by the Fourier sine inversion theorem, the relations (3.99) and (3.100) produce the solution of the pair (3.95) and (3.96) as

$$f(\tau) = \sqrt{\frac{2}{\pi}} \left[\int_0^a G_1(t) \sin \tau t \ dt + \int_a^\infty G_2(t) \sin \tau t \ dt \right]. \tag{3.102}$$

(iii) Now, we consider the more general pair

$$\int_0^\infty [1 + \omega(\tau)] f(\tau) P_{-\frac{1}{2}+i\tau}^\mu (\cosh x) d\tau = g(x), \ 0 \le x \le a, \tag{3.103}$$

$$\int_0^\infty f(\tau)\, \Gamma\left(\frac{1}{2} - \mu + i\tau\right)\, \Gamma\left(\frac{1}{2} - \mu - i\tau\right)\, \tau \sinh \pi\tau$$

$$\times\, P^\mu_{-\frac{1}{2}+i\tau}(\cosh x)\, d\tau = h(x),\quad a < x < \infty, \tag{3.104}$$

where $w(\tau)$ is a known weight function.

Equation (3.103) can be written as

$$\int_0^\infty f(\tau)\, P^\mu_{-\frac{1}{2}+i\tau}(\cosh x)d\tau = g(x) - \int_0^\infty w(\tau)\, f(\tau)\, P^\mu_{-\frac{1}{2}+i\tau}(\cosh x)d\tau,\quad 0 \le x \le a. \tag{3.105}$$

Then by the result (3.92), the solution of the pair (3.105) and (3.104) is given by

$$f(\tau) = \sqrt{\frac{2}{\pi}} \left[\int_0^a G_1(t)\, \cos \tau t\, dt + \int_a^\infty G_2(t)\, \cos \tau t\, dt\right], \tag{3.106}$$

where $G_2(t)$ is the same as defined by (3.91), but

$$G_1(t) = \pi^{-1}\, \cos \pi\mu\, \Gamma\left(\frac{1}{2} - \mu\right) \frac{d}{dt} \int_0^t \sinh^{1-\mu} x\, (\cosh t - \cosh x)^{-\frac{1}{2}+\mu}$$

$$\times \left\{g(x) - \int_0^\infty w(\tau)\, f(\tau)\, P^\mu_{-\frac{1}{2}+i\tau}(\cosh x)\, d\tau\right\} dx. \tag{3.107}$$

Let us substitute

$$G(t) = \pi^{-1} \cos \pi\mu\, \Gamma\left(\frac{1}{2} - \mu\right) \frac{d}{dt} \int_0^t \sinh^{1-\mu} x\, (\cosh t - \cosh x)^{-\frac{1}{2}+\mu}\, dx$$

and

$$\overline{H}(\tau) = \sqrt{\frac{2}{\pi}} \int_a^\infty G_2(t)\, \cos \tau t\, dt.$$

Then, using the relation (A.2.18), we find that

$$\pi^{-1}\, \Gamma\left(\frac{1}{2} - \mu\right) \cos \pi\mu\, \frac{d}{dt} \int_0^t \sinh^{1-\mu} x\, (\cosh t - \cosh x)^{-\frac{1}{2}+\mu}$$

$$\times \left\{\int_0^\infty w(\tau)\, f(\tau)\, P^\mu_{-\frac{1}{2}+i\tau}(\cosh x)\, d\tau\right\} dx$$

$$= \sqrt{\frac{2}{\pi}} \int_0^\infty w(\tau)\, f(\tau)\, \cos \tau t\, d\tau$$

$$= \frac{2}{\pi} \int_0^a G_1(s) \left\{\int_0^\infty w(\tau)\, \cos \tau t\, \cos \tau s\, d\tau\right\} ds$$

$$+ \sqrt{\frac{2}{\pi}} \int_0^\infty \overline{H}(\tau)\, w(\tau)\, \cos \tau t\, d\tau.$$

112

Thus, the solution of the pair (3.103) and (3.104) is given by the relation (3.106), where $G_1(\tau)$ is the solution of the Fredholm integral equation of the second kind

$$G_1(t) + \int_0^a K_1(t, u) \, F_1(u) \, du = G^*(t), \ 0 \le t \le a,$$

with the kernel $K_1(t, u)$ defined by

$$
\begin{aligned}
K_1(t, u) &= \frac{1}{\sqrt{2\pi}} \{w_c(t + u) + w_c(|t - u|)\}, \\
w_c(t) &= F_c[w(\tau)], \\
G^*(t) &= G(t) - \overline{G}(t), \\
\overline{G}(t) &= F_c[\overline{H}(\tau) \, w(\tau)].
\end{aligned}
$$

(iv) Lastly, the solution of the more general pair

$$\int_0^\infty \tau \, f(\tau) \, P^\mu_{-\frac{1}{2}+i\tau}(\cosh x) \, d\tau = g(x), \ 0 \le x \le a,$$

$$\int_0^\infty [1 + w(\tau)] \, f(\tau) \, \Gamma\left(\frac{1}{2} - \mu + i\tau\right) \Gamma\left(\frac{1}{2} - \mu - i\tau\right)$$

$$\times \sinh \pi\tau \, P^\mu_{-\frac{1}{2}+i\tau}(\cosh x) \, d\tau = h(x), \ a < x < \infty,$$

is given by

$$f(\tau) = \sqrt{\frac{2}{\pi}} \left[\int_0^a G_1(t) \, \sin \tau t \, dt + \int_a^\infty G_2(t) \, \sin \tau t \, dt \right],$$

where $G_1(t)$ is the same as defined by the relation (3.98) and $G_2(t)$ is the solution of the Fredholm integral equation of the second kind

$$G_2(t) = H^*(t) + \int_a^\infty K_2(t, s) \, G_2(s) \, ds,$$

where the kernel $K_2(t, s)$ is defined by

$$
\begin{aligned}
K_2(t, s) &= \frac{1}{\sqrt{2\pi}} \{w_c(t + s) - w_c(|t - s|)\}, \\
w_c(t) &= F_c[w(\tau)],
\end{aligned}
$$

113

and the function $H^*(t)$ is defined by

$$
\begin{aligned}
H^*(t) &= H_1(t) - H_2(t), \\
H_1(t) &= -\pi^{-2} \cos \pi\mu \ \Gamma\left(\frac{1}{2} + \mu\right) \frac{d}{dt} \int_t^\infty \frac{\sinh^{1+\mu} x \ h(x)}{(\cosh x - \cosh t)^{\frac{1}{2}+\mu}} \ dx, \quad a < t < \infty, \\
H_2(t) &= \frac{2}{\pi} \int_0^a G_1(s) \left\{ \int_0^\infty \omega(\tau) \ \sin \tau t \ \sin \tau s \ d\tau \right\} ds.
\end{aligned}
$$

3.2.3 Srivastava (1991) studied two pairs of integral equations involving associated Legendre functions of the first kind as kernel, which have been solved in closed form by the special method mentioned in §1.2.2.

The first pair is

$$
\int_0^\infty \left(\frac{1}{4} + \tau^2\right) f(\tau) \ P^\mu_{-\frac{1}{2}+i\tau}(\cosh x) \ d\tau = -g(x), \ 0 \le x \le a, \tag{3.108}
$$

$$
\int_0^\infty \pi^{-1} \ \tau \ \sinh \pi\tau \ \Gamma\left(\frac{1}{2} - \mu + i\tau\right) \ \Gamma\left(\frac{1}{2} - \mu - i\tau\right)
$$
$$
\times f(\tau) \ P^\mu_{-\frac{1}{2}+i\tau}(\cosh x) \ d\tau = 0, \ a \le x < \infty, \tag{3.109}
$$

To solve this pair, we assume that

$$
\int_0^\infty \pi^{-1} \ \tau \ \sinh \pi\tau \ \Gamma\left(\frac{1}{2} - \mu + i\tau\right) \ \Gamma\left(\frac{1}{2} - \mu - i\tau\right)
$$
$$
\times f(\tau) \ P^\mu_{-\frac{1}{2}+i\tau}(\cosh x) \ d\tau = \phi(x), \ 0 \le x \le a, \tag{3.110}
$$

where $\phi(x)$ is an unknown function to be determined.

From (3.109) and (3.110), it is seen that $\phi(a) = 0$ for the continuity of $\phi(x)$ at $x = a$. By the associated Mehler–Fock inversion theorem (cf. Mandal and Mandal 1997, p. 47), from the relations (3.109) and (3.110), we get

$$
f(\tau) = \int_0^a \phi(s) \ P^\mu_{-\frac{1}{2}+i\tau}(\cosh s) \ \sinh s \ ds. \tag{3.111}
$$

Substituting (3.111) into (3.108), we get

$$
\int_0^\infty \left(\frac{1}{4} + \tau^2\right) P^\mu_{-\frac{1}{2}+i\tau}(\cosh x) \ \sinh x \left\{ \int_0^a \phi(s) \ P^\mu_{-\frac{1}{2}+i\tau}(\cosh s) \ \sinh s \ ds \right\} d\tau
$$
$$
= -\sinh x \ g(x), \ 0 \le x \le a. \tag{3.112}
$$

114

Equation (3.112) is equivalent to the ODE

$$\sinh x \, \frac{d^2 u}{dx^2} + \cosh x \, \frac{du}{dx} - \frac{\mu^2 u}{\sinh x} = \sinh x \, g(x), \ 0 \le x \le a, \qquad (3.113)$$

where

$$u(x) = \int_0^\infty P^\mu_{-\frac{1}{2}+i\tau}(\cosh x) \left\{ \int_0^a \phi(s) \, P^\mu_{-\frac{1}{2}+i\tau}(\cosh s) \, \sinh s \, ds \right\} d\tau. \qquad (3.114)$$

The solution of the ODE (3.113) is

$$u(x) = \left\{ C_1 + \frac{1}{2\mu} \int_0^x \coth^\mu \frac{t}{2} \, g(t) \, \sinh t \, dt \right\} \tanh^\mu \frac{x}{2}$$

$$+ \left\{ C_2 + \frac{1}{2\mu} \int_0^x \tanh^\mu \frac{t}{2} \, g(t) \, \sinh t \, dt \right\} \coth^\mu \frac{x}{2}, \qquad (3.115)$$

when $\mu \neq 0$, and when $\mu = 0$, then

$$u(x) = C_1 + C_2 \, \log(\tanh \frac{x}{2}) + \int_0^x \operatorname{cosech} t \left\{ \int_0^t g(s) \, \sinh s \, ds \right\} dt, \qquad (3.116)$$

where C_1 and C_2 are arbitrary constants. Since $u(x)$ must be finite at $x = 0$, in (3.115) $C_2 = 0$ if $0 < \mu < \frac{1}{2}$, and $C_1 = 0$ if $-\frac{1}{2} < \mu < 0$. When $\mu = 0$, then in (3.116), $C_2 = 0$. The other constant will be determined by using the continuity condition $\phi(a) = 0$.

The equation (3.114) can be written as

$$u(x) = \int_0^a \phi(s) \, \sinh s \left\{ \int_0^\infty P^\mu_{-\frac{1}{2}+i\tau}(\cosh s) \, P^\mu_{-\frac{1}{2}+i\tau}(\cosh x) \, d\tau \right\} ds. \qquad (3.117)$$

Using the result (A.2.12), then interchanging the order of integration, the above equation (3.117) becomes

$$\left\{ \Gamma\left(\frac{1}{2} - \mu\right) \right\}^2 \int_0^x (\cosh x - \cosh t)^{-\frac{1}{2}-\mu}$$

$$\times \left\{ \int_t^a \frac{\sinh^{1+\mu} s \, \phi(s)}{(\cosh s - \cosh t)^{\frac{1}{2}+\mu}} \, ds \right\} dt = \sinh^\mu x \, u(x) = U(x), \ \text{say.} \qquad (3.118)$$

By Abel's inversion theorem, equation (3.118) produces

115

$$\int_t^a \frac{\sinh^{1+\mu} s \; \phi(s)}{(\cosh s - \cosh t)^{\frac{1}{2}+\mu}} \, ds$$

$$= \frac{\Gamma(\frac{1}{2}-\mu)}{\Gamma(\frac{1}{2}+\mu)} \sinh t \left[\frac{U(0)}{(\cosh t - 1)^{\frac{1}{2}-\mu}} \right.$$

$$\left. + \int_0^t \frac{U'(x) \, dx}{(\cosh t - \cosh x)^{\frac{1}{2}-\mu}} = \psi(t), \quad \text{say.} \right. \tag{3.119}$$

By another use of Abel's inversion theorem, equation (3.119) gives

$$\phi(s) = \frac{\sinh^{-\mu} s}{\left\{ \Gamma(\frac{1}{2}+\mu) \right\}^2} \int_0^a \frac{\psi'(y) \, dy}{(\cosh y - \cosh s)^{\frac{1}{2}-\mu}}, \tag{3.120}$$

where $\psi(a) = 0$. The other constant is determined from the condition $\psi(a) = 0$ arising from the requirement of continuity of $\phi(s)$ at $s = a$ and hence the constant is determined by the relation

$$\int_0^a \frac{U'(x) \, dx}{(\cosh a - \cosh x)^{\frac{1}{2}-\mu}} + \frac{U(0)}{(\cosh a - 1)^{\frac{1}{2}-\mu}} = 0. \tag{3.121}$$

Again, the second pair is

$$\int_0^\infty \tau \, f(\tau) \, P_{-\frac{1}{2}+i\tau}^{\mu}(\cosh x) \, d\tau = 0, \; 0 \leq x \leq a, \tag{3.122}$$

$$\int_0^\infty \pi^{-1} \left(\frac{1}{4} + \tau^2 \right) \sinh \pi\tau \; \Gamma \left(\frac{1}{2} - \mu + i\tau \right) \Gamma \left(\frac{1}{2} - \mu - i\tau \right)$$

$$\times f(\tau) \, P_{-\frac{1}{2}+i\tau}^{\mu}(\cosh x) \, d\tau = -g(x), \; a \leq x < \infty. \tag{3.123}$$

In this case, let us assume that

$$\int_0^\infty \tau \, f(\tau) \, P_{-\frac{1}{2}+i\tau}^{\mu}(\cosh x) \, d\tau = \phi(x), \; a \leq x < \infty, \tag{3.124}$$

where $\phi(x)$ is an unknown function to be determined. Then by the associated Mehler−Fock inversion theorem, equations (3.122) and (3.124) produce

$$f(\tau) = \pi^{-1} \, \sinh \pi\tau \; \Gamma \left(\frac{1}{2} - \mu + i\tau \right) \Gamma \left(\frac{1}{2} - \mu - i\tau \right)$$

$$\times \int_a^\infty \phi(s) \, P_{-\frac{1}{2}+i\tau}^{\mu}(\cosh s) \, \sinh s \, ds. \tag{3.125}$$

116

Equation (3.123) is equivalent to the ODE

$$\sinh x \, \frac{d^2u}{dx^2} + \cosh x \, \frac{du}{dx} - \frac{\mu^2 \, u}{\sinh x} = \sinh x \, g(x), \quad a \le x < \infty, \tag{3.126}$$

where

$$u(x) = \int_0^\infty \pi^{-1} \sinh \pi\tau \, \Gamma(\tfrac{1}{2} - \mu + i\tau) \, \Gamma(\tfrac{1}{2} - \mu - i\tau) \, f(\tau) \, P^\mu_{-\frac{1}{2}+i\tau} (\cosh x) \, d\tau. \tag{3.127}$$

Substituting the vlaue of $f(\tau)$ from (3.125) into the relation (3.127), interchanging the order of integration, then applying the result (A.2.13) and again interchanging the order of integration, we get

$$\left\{ \Gamma\left(\tfrac{1}{2} + \mu\right) \right\}^{-2} \int_x^\infty (\cosh t - \cosh x)^{-\frac{1}{2}+\mu} \left\{ \int_a^t \frac{\sinh^{1-\mu} s \, \phi(s)}{(\cosh t - \cosh s)^{\frac{1}{2}-\mu}} \, ds \right\} dt$$

$$= \sinh^\mu x \, u(x) = U(x), \quad \text{say.} \tag{3.128}$$

The solution of the ODE (3.126) has already been obtained earlier by the relations (3.115) and (3.116). The solution should be such that $u(x) \to 0$ as $x \to \infty$. Therefore, for $-\tfrac{1}{2} < \mu < \tfrac{1}{2}$, $C_2 = -C_1$ and for $\mu = 0$, $C_1 = 0$.

By Abel's inversion theorem, (3.128) becomes

$$\int_a^t \frac{\sinh^{1-\mu} s \, \phi(s)}{(\cosh t - \cosh s)^{\frac{1}{2}-\mu}} \, ds = -\frac{\Gamma(\tfrac{1}{2} + \mu)}{\Gamma(\tfrac{1}{2} - \mu)} \sinh t \int_t^\infty \frac{U'(x) \, dx}{(\cosh x - \cosh t)^{\frac{1}{2}+\mu}}$$

$$= \psi(t), \quad \text{say.} \tag{3.129}$$

By another use of Abel's inversion theorem, the solution of the integral equation (3.129) is obtained as

$$\phi(s) = \left\{ \Gamma\left(\tfrac{1}{2} - \mu\right) \right\}^2 \sinh^\mu s \int_a^s \frac{\psi'(t) \, dt}{(\cosh s - \cosh t)^{\frac{1}{2}+\mu}}. \tag{3.130}$$

The continuity requirement for the function $\phi(x)$ at $x = a$ gives $\phi(a) = 0$ for which $\psi(a) = 0$. Thus, the arbitrary constant is determined by the relation

$$\int_a^\infty \frac{U'(x) \, dx}{(\cosh x - \cosh a)^{\frac{1}{2}+\mu}} = 0.$$

117

Thus, the explicit solution of the pair (3.122) and (3.123) is obtained by the relations (3.125) and (3.130).

3.2.4 Mandal (1992) considered a pair of integral equations of the form

$$\int_0^\infty \tau^{-1} f(\tau) \, \tanh c\tau \, P^\mu_{-\frac{1}{2}+i\tau}(\cosh x) \, d\tau = g(x), \ 0 < x < a, \qquad (3.131)$$

$$\int_0^\infty f(\tau) \, P^\mu_{-\frac{1}{2}+i\tau}(\cosh x) \, d\tau = h(x), \ x > a, \qquad (3.132)$$

where $Re \ \mu < \frac{1}{2}$. An explicit solution to this pair is obtained. For $\mu = 0$, $h(x) = 0$, the corresponding pair is given by (3.20) and (3.21) whose solution is obtained in Section 3.1. To obtain the solution of this pair, we first find an explicit solution to the pair (3.131) and (3.132) with $h(x) = 0$, and then an explicit solution to the pair with $g(x) = 0$. The desired solution of the dual integral equations (3.131) and (3.132) is then obtained by combining the solutions of these two pairs.

(A) $h(x) = 0$.

In this case, we consider the dual integral equations

$$\int_0^\infty \tau^{-1} A(\tau) \, P^\mu_{-\frac{1}{2}+i\tau}(\cosh x) \, \tanh c\tau \, d\tau = g(x), \ 0 < x < a, \qquad (3.133)$$

$$\int_0^\infty A(\tau) \, P^\mu_{-\frac{1}{2}+i\tau}(\cosh x) \, d\tau = 0, \ x > a, \qquad (3.134)$$

where $Re \ \mu < \frac{1}{2}$.

Multiplying (3.133) by

$$\pi^{-1} \, \Gamma\left(\frac{1}{2} - \mu\right) \, \cos \pi\mu \, \sinh^{1-\mu} x \, (\cosh s - \cosh x)^{\mu-\frac{1}{2}},$$

integrating the relation with respect to x from 0 to s $(< a)$, and then differentiating both sides with respect to s, we find after utilizing the result (A.2.18), that

$$\int_0^\infty \tau^{-1} A(\tau) \, \tanh c\tau \, \cos \tau s \, d\tau = F(s), \ 0 < s < a, \qquad (3.135)$$

where

$$F(s) = \frac{\Gamma(\frac{1}{2} - \mu) \cos \pi\mu}{\sqrt{2\pi}} \, \frac{d}{ds} \int_0^s \frac{\sinh^{1-\mu} x \, g(x)}{(\cosh s - \cosh x)^{1-\mu}} \, dx. \qquad (3.136)$$

118

Let us assume that

$$A(\tau) = \int_0^a \phi(t) \cos \tau t \, dt, \qquad (3.137)$$

where $\phi(t)$ is an unknown function to be determined. If we substitute (3.137) into (3.135) and interchange the order of integration, we find that

$$\frac{1}{2} \int_0^a \phi(t) \, \log \left| \frac{\cosh \frac{\pi s}{2c} + \cosh \frac{\pi t}{2c}}{\cosh \frac{\pi s}{2c} - \cosh \frac{\pi t}{2c}} \right| \, dt = F(s), \ 0 < s < a,$$

which after differentiation with respect to s produces a singular integral equation for $\phi(t)$ given by

$$\int_0^a \frac{\cosh \frac{\pi t}{2c} \, \phi(t)}{\cosh \frac{\pi t}{c} - \cosh \frac{\pi s}{c}} \, dt = \frac{c}{\pi} \frac{F'(s)}{\sinh \frac{\pi s}{2c}}, \ 0 < s < a, \qquad (3.138)$$

where the integral is in the sense of the Cauchy principal value. Substituting

$$\left. \begin{aligned}
\xi &= \cosh \frac{\pi t}{c}, \ \eta = \cosh \frac{\pi s}{c}, \ k = \cosh \frac{\pi a}{c}, \\[2mm]
\Phi(\xi) &= \frac{\phi(t)}{\sinh \frac{\pi t}{2c}}, \ L(\eta) = -\frac{2}{\pi} \frac{F'(s)}{\sinh \frac{\pi s}{2c}},
\end{aligned} \right\} \qquad (3.139)$$

the above integral equation become

$$\int_1^k \frac{\Phi(\xi)}{\eta - \xi} \, d\xi = \pi \, L(\eta), \ 1 < \eta < k. \qquad (3.140)$$

One form of the solution of this integral equation is (cf. Cooke 1970)

$$\Phi(\xi) = \frac{1}{\pi} \left(\frac{\xi - 1}{k - \xi} \right)^{\frac{1}{2}} \int_1^k \left(\frac{k - \eta}{k + \eta} \right)^{\frac{1}{2}} \frac{L(\eta)}{\eta - \xi} \, d\eta + \frac{A}{\sqrt{(\xi - 1)(k - \xi)}},$$

where A is an arbitrary constant. The back substitution from (3.139) produces the desired solution in the form

$$\phi(t) = -\frac{4\sqrt{\pi}}{\pi c} \, \chi(t) + \frac{A}{\sqrt{2}} \left(\cosh \frac{\pi a}{c} - \cosh \frac{\pi t}{c} \right)^{-\frac{1}{2}}, \qquad (3.141)$$

where

$$\chi(t) = \frac{\sinh \frac{2\pi t}{c}}{\sqrt{\cosh \frac{\pi a}{c} - \cosh \frac{\pi t}{c}}} \int_0^a \sqrt{\frac{\cosh \frac{\pi a}{c} - \cosh \frac{\pi s}{c}}{\cosh \frac{\pi s}{c} - 1}} \frac{\cosh \frac{\pi s}{2c} \, F'(s)}{\cosh \frac{\pi s}{c} - \cosh \frac{\pi t}{c}} \, ds. \qquad (3.142)$$

119

To find the arbitrary constant A in (3.141), we use (3.137) in equation (3.134) and utilizing the result (A.2.10) we find that

$$\int_0^a \frac{\phi(t)\, dt}{(\cosh x - \cosh t)^{\frac{1}{2}+\mu}} = 0, \ x > a. \tag{3.143}$$

If we multiply both sides of (3.143) by $\sinh \frac{x}{2} \cosh^{2\mu-2} \frac{x}{2}$ and integrate the relation with respect to x from a to ∞, we obtain

$$2^{\frac{1}{2}-\mu} \int_0^a \frac{\phi(t)}{\cosh^2 \frac{t}{2}} \, dt - \cosh^{2\mu-1} \frac{a}{2} \int_0^a \frac{(\cosh a - \cosh t)^{\frac{1}{2}-\mu}}{\cosh^2 \frac{t}{2}} \phi(t)\, dt = 0. \tag{3.144}$$

Substituting the value for $\phi(t)$ from (3.141) into (3.144), a linear equation for A is obtained as given by

$$A = \frac{J_1 - J_2}{I_1 - I_2}, \tag{3.145}$$

where

$$I_1 = 2^{\frac{1}{2}-\mu} \int_0^a \frac{dt}{\cosh^2 \frac{t}{2} (\cosh \frac{\pi a}{c} - \cosh \frac{\pi t}{c})^{\frac{1}{2}}},$$

$$I_2 = \cosh^{2\mu-1} \frac{a}{2} \int_0^a \frac{(\cosh a - \cosh t)^{\frac{1}{2}-\mu}}{\cosh^2 \frac{t}{2} (\cosh \frac{\pi a}{c} - \cosh \frac{\pi t}{c})^{\frac{1}{2}}} \, dt,$$

$$J_1 = \frac{8}{\pi c} 2^{\frac{1}{2}-\mu} \int_0^a \frac{\chi(t)}{\cosh^2 \frac{t}{2}} \, dt,$$

and

$$J_2 = \frac{8}{\pi c} \cosh^{2\mu-1} \frac{a}{2} \int_0^a \frac{(\cosh a - \cosh t)^{\frac{1}{2}-\mu}}{\cosh^2 \frac{t}{2}} \chi(t)\, dt.$$

Since $\phi(t)$ is now known completely, an explicit solution of the pair (3.133) and (3.134) is obtained by the relation (3.137).

(B) $g(x) = 0.$

In this case, we consider the dual integral equations

$$\int_0^\infty \tau^{-1} B(\tau) \tanh c\tau \, P^\mu_{-\frac{1}{2}+i\tau} (\cosh x) \, d\tau = 0, \ 0 < x < a, \tag{3.146}$$

$$\int_0^\infty B(\tau) P^\mu_{-\frac{1}{2}+i\tau} (\cosh x) \, d\tau = h(x), \ x > a, \tag{3.147}$$

120

where $Re\ \mu < \frac{1}{2}$. By using a procedure similar to that of Case (A), the equation (3.146) reduces to

$$\int_0^\infty \tau^{-1}\ B(\tau)\ \tanh c\tau\ \cos \tau s\ d\tau = 0,\ 0 < s < a. \tag{3.148}$$

Let us assume as before

$$B(\tau) = \int_0^a \psi(t)\ \cos \tau t\ dt. \tag{3.149}$$

Then (3.147) gives after utilizing the result (A.2.10) that

$$\int_0^a \frac{\psi(t)\ dt}{(\cosh x - \cosh t)^{\frac{1}{2}+\mu}} = H(x),\ x > a, \tag{3.150}$$

where

$$H(x) = \sqrt{\frac{2}{\pi}}\ \Gamma\left(\frac{1}{2} - \mu\right)\ \sinh^{-\mu} x\ h(x).$$

Using the equations (3.149) in (3.148), we find that $\chi(t)$ satisfies the homogeneous singular integral equation

$$\int_0^a \frac{\cosh \frac{\pi t}{2c}\ \psi(t)}{\cosh \frac{\pi t}{c} - \cosh \frac{\pi s}{c}}\ dt = 0,\ 0 < s < a.$$

Its solution is given by

$$\psi(t) = \frac{B}{\sqrt{2}}\left(\cosh \frac{\pi a}{c} - \cosh \frac{\pi t}{c}\right)^{-\frac{1}{2}},$$

where B is an arbitrary constant which has to be found from (3.150). For this, we multiply both sides of (3.150) by $\sinh \frac{x}{2} \cosh^{2\mu-2} \frac{x}{2}$ and integrate with respect to x from a to ∞ to obtain

$$2^{\frac{1}{2}-\mu} \int_0^a \frac{\psi(t)\ dt}{\cosh^2 \frac{t}{2}} - \cosh^{2\mu-1} \frac{a}{2} \int_0^a \frac{(\cosh a - \cosh t)^{\frac{1}{2}-\mu}}{\cosh^2 \frac{t}{2}}\ \psi(t)\ dt$$

$$= (1 - 2\mu) \int_a^\infty \frac{\sinh \frac{x}{2}\ H(x)}{\cosh^{2-2\mu} \frac{x}{2}}\ dx = I,\ \text{say}.$$

Substituting the value of $\psi(t)$ in the above relation, we find a linear relation for B as

$$B = \frac{J}{I_1 - I_2},$$

121

where I_1 and I_2 are given by the relations as obtained in Case (A).

Thus $\psi(t)$ is now completely determined so that an explicit solution to the dual integral equations (3.146) and (3.147) is obtained by the relation (3.149). The solution of the pair (3.131) and (3.132) can now be obtained in closed form by adding the solutions $A(\tau)$ and $B(\tau)$.

3.2.5 Now we consider some dual integral equations involving the associated Legendre function $P^{\mu}_{-\frac{1}{2}+i\tau}(\cosh x)$ with $-\frac{1}{2} < Re\ \mu \leq 0$ studied by Mandal and Mandal (1993). The solutions of these pairs of integral equations are obtained by the special method mentioned in §1.2.2. The pairs of dual integral equations are the following.

(A) The first pair is

$$\int_0^\infty f(\tau)\ P^{\mu}_{-\frac{1}{2}+i\tau}(\cosh x)\ d\tau = g(x),\ 0 \leq x \leq a, \tag{3.151}$$

$$\int_0^\infty \tau\ \Gamma\left(\frac{1}{2} - \mu + i\tau\right)\ \Gamma\left(\frac{1}{2} - \mu - i\tau\right)\ \sinh \pi\tau$$

$$\times f(\tau)\ P^{\mu}_{-\frac{1}{2}+i\tau}(\cosh x)\ d\tau = 0,\ a \leq x < \infty, \tag{3.152}$$

where $-\frac{1}{2} < Re\ \mu \leq 0$.

To solve these dual integral equations, we assume the right side of equation (3.152) to be equal to an unknown function $\phi(x)$ for $0 \leq x \leq a$, so that by applying the associated Mehler−Fock inversion formulae mentioned earlier, we find

$$\pi\ f(\tau) = \int_0^a \phi(s)\ P^{\mu}_{-\frac{1}{2}+i\tau}(\cosh s)\ \sinh s\ ds. \tag{3.153}$$

Using the above equation (3.153), then interchanging the order of integration and using the result (A.2.12) in (3.151), we obtain Abel's integral equation

$$\int_0^x \frac{dt}{(\cosh x - \cosh t)^{\frac{1}{2}+\mu}} \int_t^a \frac{\sinh^{1+\mu} s\ \phi(s)}{(\cosh s - \cosh t)^{\frac{1}{2}+\mu}}\ ds$$

$$= \frac{\pi\left\{\Gamma(\frac{1}{2}+\mu)\right\}^2\ g(x)}{\sinh^\mu x},\ 0 \leq x \leq a,\ -\frac{1}{2} < Re\ \mu \leq 0, \tag{3.154}$$

122

so that

$$\int_t^a \frac{\sinh^{1+\mu} s \; \phi(s)}{(\cosh s - \cosh t)^{\frac{1}{2}+\mu}} = \cos \pi\mu \; \frac{d}{dt} \int_0^t \frac{\left\{ \Gamma(\frac{1}{2} - \mu) \right\}^2 \sinh^{1-\mu} u \; g(u)}{(\cosh t - \cosh u)^{\frac{1}{2}-\mu}} \; dt$$

$$= \psi(t), \; 0 \le t \le a. \tag{3.155}$$

Another use of Abel's inversion theorem gives

$$\sinh^{1+\mu} x \; \phi(x) = -\frac{\cos \pi\mu}{\pi} \int_x^a \frac{\sinh t \; \psi(t) \; dt}{(\cosh t - \cosh x)^{\frac{1}{2}-\mu}}. \tag{3.156}$$

Hence, the solution of the pair (3.151) and (3.152) is obtained, after some elementary calculations and using the result (A.2.18) as

$$f(\tau) = \frac{\sqrt{2}}{\pi\sqrt{\pi}} \; \Gamma\left(\frac{1}{2} - \mu\right) \; \cos \pi\mu \int_0^a \cos t\tau \left\{ \frac{d}{dt} \int_0^t \frac{\sinh^{1-\mu} x \; g(x)}{(\cosh t - \cosh x)^{\frac{1}{2}-\mu}} \; dx \right\} dt, \tag{3.157}$$

where $-\frac{1}{2} < Re \; \mu \le 0$. This result can also be obtained by using the result (3.92) derived earlier by another method. For $\mu = 0$ this also reduces to the known result obtained by Babloian (1964).

(B) Next we consider the pair

$$\int_0^\infty \tau^2 \; f(\tau) \; P_{-\frac{1}{2}+i\tau}^\mu(\cosh x) \; d\tau = g(x), \; 0 \le x \le a, \tag{3.158}$$

$$\int_0^\infty \tau \; \Gamma\left(\frac{1}{2} - \mu + i\tau\right) \; \Gamma\left(\frac{1}{2} - \mu - i\tau\right) \; \sinh \pi\tau \; f(\tau)$$

$$\times P_{-\frac{1}{2}+i\tau}^\mu (\cosh x) \; d\tau = 0, \; a \le x < \infty, \; -\frac{1}{2} < Re \; \mu \le 0. \tag{3.159}$$

As before, we assume the right side of equation (3.159) to be an unknown function $\phi(x)$ for $0 \le x \le a$. Then by the asociated Mehler–Fock inversion theorem, we obtain

$$\pi \; f(\tau) = \int_0^a \phi(x) \; P_{-\frac{1}{2}+i\tau}^\mu(\cosh x) \; \sinh x \; dx. \tag{3.160}$$

Substituting $f(\tau)$ in equation (3.158), we have

$$\int_0^\infty \tau^2 \, P_{-\frac{1}{2}+i\tau}^\mu(\cosh x) \left\{ \int_0^a \phi(s) \, P_{-\frac{1}{2}+i\tau}^\mu(\cosh s) \, \sinh s \, ds \right\} d\tau$$

$$= \pi \, g(x), \ 0 \le x \le a. \tag{3.161}$$

This is equivalent to the ODE

$$\frac{1}{\sinh x} \frac{d}{dx} \left(\sinh x \, \frac{du}{dx} \right) + \left(\frac{1}{4} - \frac{\mu^2}{\sinh^2 x} \right) u = -\pi g(x), \ 0 \le x \le a, \tag{3.162}$$

where

$$u(x) = \int_0^\infty P_{-\frac{1}{2}+i\tau}^\mu(\cosh x) \left\{ \int_0^a \phi(s) \, P_{-\frac{1}{2}+i\tau}^\mu(\cosh s) \, \sinh s \, ds \right\} d\tau. \tag{3.163}$$

The solution of the ODE (3.162) is

$$u(x) = C_1 \, P_{-\frac{1}{2}}^\mu(\cosh x) + C_2 \, Q_{-\frac{1}{2}}^\mu(\cosh x) + 2^{-2\mu} \, e^{i\pi\mu} \frac{\Gamma\left(\frac{3}{4} - \frac{\mu}{2}\right) \Gamma\left(\frac{1}{4} - \frac{\mu}{2}\right)}{\Gamma\left(\frac{3}{4} + \frac{\mu}{2}\right) \Gamma\left(\frac{1}{4} + \frac{\mu}{2}\right)}$$

$$\times \left\{ P_{-\frac{1}{2}}^\mu(\cosh x) \int_0^x g(t) \, Q_{-\frac{1}{2}}^\mu(\cosh t) \, \sinh t \, dt \right.$$

$$\left. - Q_{-\frac{1}{2}}^\mu(\cosh x) \int_0^x g(t) \, P_{-\frac{1}{2}}^\mu(\cosh t) \sinh t \, dt \right\}, \ 0 \le x \le a, \tag{3.164}$$

where C_1 and C_2 are arbitrary constants to be determined from physical considerations of the problems. Since $u(x)$ must be bounded at $x = 0$ and $-\frac{1}{2} < Re \ \mu \le 0$, $C_2 = 0$. The constant C_1 will be determined later.

As before, (3.163) gives rise to

$$\int_0^x \frac{dt}{(\cosh x - \cosh t)^{\frac{1}{2}+\mu}} \int_t^a \frac{\sinh^{1+\mu} s \, \phi(s)}{(\cosh s - \cosh t)^{\frac{1}{2}+\mu}} ds = \frac{\left\{ \Gamma(\frac{1}{2} - \mu) \right\}^2 u(x)}{\sinh^\mu x}, \tag{3.165}$$

from which we find another Abel integral equation

$$\int_x^a \frac{\sinh^{1+\mu} s \, \phi(s)}{(\cosh s - \cosh t)^{\frac{1}{2}+\mu}} ds = \frac{\Gamma(\frac{1}{2} - \mu)}{\Gamma(\frac{1}{2} + \mu)} \frac{d}{dx} \int_0^x \frac{\sinh^{1-\mu} t \, u(t)}{(\cosh x - \cosh t)^{\frac{1}{2}-\mu}} dt$$

$$= \psi(x), \ \text{say}, \ 0 \le x \le a, \ -\frac{1}{2} < Re \ \mu \le 0,$$

so that

$$\sinh^{1+\mu} x \, \phi(x) = -\frac{\cos \pi\mu}{\pi} \frac{d}{dx} \int_x^a \frac{\sinh t \, \psi(t)}{(\cosh t - \cosh x)^{\frac{1}{2}-\mu}} dt. \tag{3.166}$$

124

This gives the complete solution of the pair (3.158) and (3.159) provided the constant C_1 is determined. This constant C_1 is obtained from the fact that $\psi(a) = 0$ if $\phi(x)$ is continuous at $x = a$. Therefore, the constant C_1 is determined from the relation

$$\int_0^a \frac{\frac{d}{dt}\left\{ u(t)\ \sinh^{-\mu} t \right\}}{(\cosh a - \cosh t)^{\frac{1}{2} - \mu}}\ dt = 0. \tag{3.167}$$

For $\mu = 0$, the corresponding dual integral equations have been considered in §3.1.2.

(C) Now we consider the dual integral equations

$$\int_0^\infty (\tau^2 - \nu^2)\ f(\tau)\ P^\mu_{-\frac{1}{2}+i\tau}(\cosh x)\ d\tau = g(x),\ 0 \le x \le a, \tag{3.168}$$

$$\int_0^\infty \tau\ \sinh \pi\tau\ \Gamma\left(\frac{1}{2} - \mu + i\tau\right)\ \Gamma\left(\frac{1}{2} - \mu - i\tau\right)\ f(\tau)$$

$$\times\ P^\mu_{-\frac{1}{2}+i\tau}(\cosh x)\ d\tau = 0,\ a \le x < \infty, \tag{3.169}$$

where $-\frac{1}{2} < Re\ \mu \le 0$ and ν is real.

As before, we assume the right side of (3.159) to be an unknown function $\phi(x)$ for $0 \le x \le a$. By using the associated Mehler$-$Fock inversion theorem, we find that

$$\pi\ f(\tau) = \int_0^a \phi(s)\ P^\mu_{-\frac{1}{2}+i\tau}(\cosh s)\ \sinh s\ ds.$$

Substituting this into (3.168), we get

$$\int_0^\infty (\tau^2 - \nu^2)\ P^\mu_{-\frac{1}{2}+i\tau}(\cosh x) \left\{ \int_0^a \phi(s)\ P^\mu_{-\frac{1}{2}+i\tau}(\cosh s)\ \sinh s\ ds \right\} d\tau$$

$$= \pi\ g(x),\ 0 \le x \le a. \tag{3.170}$$

This is equivalent to the ODE

$$\frac{1}{\sinh x}\frac{d}{dx}\left(\sinh x\ \frac{du}{dx}\right) + \left(\frac{1}{4} + \nu^2 - \frac{\mu^2}{\sinh^2 x}\right) u = -\pi\ g(x), \tag{3.171}$$

where now

$$u(x) = \int_0^\infty P^\mu_{-\frac{1}{2}+i\tau}(\cosh x) \left\{ \int_0^a \phi(s)\ P^\mu_{-\frac{1}{2}+i\tau}(\cosh s)\ \sinh s\ ds \right\} d\tau,\ 0 \le x \le a. \tag{3.172}$$

The solution of the ODE (3.171) is

$$
\begin{aligned}
u(x) \;=\; & C_1\, P^{\mu}_{-\frac{1}{2}+i\nu}(\cosh x) + C_2\, Q^{\mu}_{-\frac{1}{2}+i\nu}(\cosh x) \\
& + 2^{-2\mu}\, e^{-i\pi\mu}\, \frac{\Gamma\left(\frac{3}{4}+\frac{i\nu}{2}-\frac{\mu}{2}\right)\Gamma\left(\frac{1}{4}+\frac{i\nu}{2}-\frac{\mu}{2}\right)}{\Gamma\left(\frac{3}{4}+\frac{i\nu}{2}+\frac{\mu}{2}\right)\Gamma\left(\frac{1}{4}+\frac{i\nu}{2}+\frac{\mu}{2}\right)} \\
& \times \left\{ P^{\mu}_{-\frac{1}{2}+i\nu}(\cosh x) \int_0^x g(t)\, Q^{\mu}_{-\frac{1}{2}+i\nu}(\cosh t)\, \sinh t\, dt \right. \\
& \left. - Q^{\mu}_{-\frac{1}{2}+i\nu}(\cosh x) \int_0^x g(t)\, P^{\mu}_{-\frac{1}{2}+i\nu}(\cosh t)\sinh t\, dt \right\}. \quad (3.173)
\end{aligned}
$$

Since $u(x)$ must be fintie at $x=0$ and $-\frac{1}{2} < Re\ \mu \le 0$, the constant $C_2 = 0$. Now proceeding as in earlier cases, we obtain $\phi(x)$ for $0 \le x \le a$ and the constant C_1 by the equations (3.166) and (3.167) respectively with $u(x)$ given by the relation (3.173). This gives the complete solution of the pair (3.168) and (3.169).

(D) Finally, we consider the pair

$$
\int_0^\infty \left\{ \Gamma\left(\frac{1}{2}-\mu+i\tau\right)\Gamma\left(\frac{1}{2}-\mu-i\tau\right) \right\}^{-1} \operatorname{cosech} \pi\tau
$$

$$
\times f(\tau)\, P^{\mu}_{-\frac{1}{2}+i\tau}(\cosh x)\, d\tau = g(x), \; 0 \le x \le a, \quad (3.174)
$$

$$
\int_0^\infty \left[\lambda_1\,\tau + \mu_1(\tau^2+c^2)\left\{ \Gamma\left(\frac{1}{2}-\mu+i\tau\right)\Gamma\left(\frac{1}{2}-\mu-i\tau\right) \right\}^{-1} \operatorname{cosech} \pi\tau \right]
$$

$$
\times f(\tau)\, P^{\mu}_{-\frac{1}{2}+i\tau}(\cosh x)\, d\tau = 0, \; a \le x < \infty, \quad (3.175)
$$

where $-\frac{1}{2} < Re\ \mu \le 0$, λ_1, μ_1, c are real and $c > \frac{1}{2}$.

For $\mu_1 = 0$, this pair reduces to the pair (3.32) and (3.33) by redefining $f(\tau)$ appropriately.

When $\lambda_1 = 0$ but $\mu_1 \ne 0$, the solution to the dual integral equations can be obtained as before and is given by (after utilizing the result (1) on p. 169 of Erdélyi et al. (1953) involving associated Legendre functions in the simplification)

$$
\begin{aligned}
\pi\, f(\tau) \;=\; & \tau\, \sinh^2 \pi\tau \left\{ \Gamma\left(\frac{1}{2}-\mu+i\tau\right)\Gamma\left(\frac{1}{2}-\mu-i\tau\right) \right\}^2 \\
& \times \left[\frac{g(a)}{(\tau^2+c^2)} \left\{ (i\tau-c)\cosh a\, P^{\mu}_{-\frac{1}{2}+i\tau}(\cosh a)\, Q^{\mu}_{c-\frac{1}{2}}(\cosh a) \right. \right.
\end{aligned}
$$

$$+(\mu + c - \frac{1}{2})\, P^\mu_{-\frac{1}{2}+i\tau}(\cosh a)\, Q^\mu_{c-\frac{3}{2}}(\cosh a)$$

$$-(\mu + i\tau - \frac{1}{2})\, P^\mu_{-\frac{3}{2}+i\tau}(\cosh a)\, Q^\mu_{c-\frac{1}{2}}(\cosh a)\Big\}$$

$$+ \int_0^a g(x)\, P^\mu_{-\frac{1}{2}+i\tau}(\cosh x)\, \sinh x\, dx\Bigg].$$

(3.176)

When both λ_1 and μ_1 are not zero, we can write (3.175) as

$$\int_0^\infty (\tau^2 + c^2)\left\{\Gamma\left(\frac{1}{2} - \mu + i\tau\right)\Gamma\left(\frac{1}{2} - \mu - i\tau\right)\right\}^{-1} \operatorname{cosech} \pi\tau$$

$$\times \left\{1 + \frac{\lambda_1 \tau}{\mu(\tau^2 + c^2)}\, \sinh \pi\tau\, \Gamma\left(\frac{1}{2} - \mu + i\tau\right)\Gamma\left(\frac{1}{2} - \mu - i\tau\right)\right\}$$

$$\times f(\tau)\, P^\mu_{-\frac{1}{2}+i\tau}(\cosh x)\, d\tau = 0,\ a \le x < \infty.$$

(3.177)

As before, equation (3.177) is equivalent to the ODE

$$\frac{1}{\sinh x}\frac{d}{dx}\left(\sinh x\frac{du}{dx}\right) + \left(\frac{1}{4} - c^2 - \frac{\mu^2}{\sinh^2 x}\right)u = 0,$$

(3.178)

where

$$u(x) = \int_0^\infty \left[1 + \frac{\lambda_1 \tau}{\mu(\tau^2 + c^2)}\, \sinh \pi\tau\, \Gamma\left(\frac{1}{2} - \mu + i\tau\right)\Gamma\left(\frac{1}{2} - \mu - i\tau\right)\right]$$

$$\times \left\{\Gamma\left(\frac{1}{2} - \mu + i\tau\right)\Gamma\left(\frac{1}{2} - \mu - i\tau\right)\right\}^{-1}$$

$$\times \operatorname{cosech} \pi\tau\, f(\tau)\, P^\mu_{-\frac{1}{2}+i\tau}(\cosh x)\, d\tau,\ a \le x < \infty.$$

(3.179)

The solution of the ODE (3.178) is

$$u(x) = C_1\, P^\mu_{c-\frac{1}{2}}(\cosh x) + C_2\, Q^\mu_{c-\frac{1}{2}}(\cosh x),\ a \le x < \infty.$$

Since $u(x)$ must be finite as $x \to \infty$ and $c > \frac{1}{2}$, we must have $C_1 = 0$. Hence, by (3.179) we get

$$\int_0^\infty \left\{\Gamma\left(\frac{1}{2} - \mu + i\tau\right)\Gamma\left(\frac{1}{2} - \mu - i\tau\right)\right\}^{-1} \operatorname{cosech} \pi\tau\, f(\tau)\, P^\mu_{-\frac{1}{2}+i\tau}(\cosh x)\, d\tau$$

$$= C_2\, Q^\mu_{c-\frac{1}{2}}(\cosh x) - \frac{\lambda_1}{\mu_1}\int_0^\infty \frac{t\, f(t)}{(t^2 + c^2)}\, P^\mu_{-\frac{1}{2}+it}(\cosh x)\, dt,\ a \le x < \infty.$$

(3.180)

127

Equations (3.180) and (3.174) give, after using the associated Mehler–Fock inversion theorem,

$$\pi \left\{ \Gamma\left(\frac{1}{2} - \mu + i\tau\right) \Gamma\left(\frac{1}{2} - \mu - i\tau\right) \right\}^{-2} \operatorname{cosech}^2 \pi\tau \, \frac{f(\tau)}{\tau}$$

$$
\begin{aligned}
= \; & C_2 \int_a^\infty P^\mu_{-\frac{1}{2}+i\tau}(\cosh x) \, Q^\mu_{c-\frac{1}{2}}(\cosh x) \, \sinh x \, dx \\
& + \int_0^a g(x) \, P^\mu_{-\frac{1}{2}+i\tau}(\cosh x) \, \sinh x \, dx \\
& - \frac{\lambda_1}{\mu_1} \int_a^\infty P^\mu_{-\frac{1}{2}+i\tau}(\cosh x) \, \sinh x \\
& \times \left\{ \int_0^\infty \frac{t \, f(t)}{(t^2 + c^2)} P^\mu_{-\frac{1}{2}+it}(\cosh x) \, dt \right\} dx.
\end{aligned}
\tag{3.181}
$$

The continuity requirement at $x = a$ gives

$$C_2 = \frac{g(a)}{Q^\mu_{c-\frac{1}{2}}(\cosh a)} + \frac{\lambda_1}{\mu_1 \, Q^\mu_{c-\frac{1}{2}}(\cosh a)} \int_0^\infty \frac{t \, f(t)}{(t^2 + c^2)} \, P^\mu_{-\frac{1}{2}+it}(\cosh a) \, dt.$$

Substituting the value of C_2 in (3.181) and after some simplification, we find the Fredholm integral equation of the second kind for $f(\tau)$ as

$$\frac{f(\tau)}{\tau} - \lambda \int_0^\infty \frac{f(t)}{t} \, K(t, \tau) \, dt = B(\tau),
\tag{3.182}$$

where

$$
\begin{aligned}
K(t, \tau) = \; & \frac{t^2 \, (t^2 + c^2)^{-1} \left\{ \Gamma(\frac{1}{2} - \mu + i\tau) \, \Gamma(\frac{1}{2} - \mu - i\tau) \right\}^2 \sinh^2 \pi\tau}{\pi \left\{ \lambda_1 \, \tau \, \Gamma(\frac{1}{2} - \mu + i\tau) \, \Gamma(\frac{1}{2} - \mu - i\tau) \, \sinh \pi\tau + \mu_1 \, (\tau^2 + c^2) \right\}} \\
& \times \left[\frac{P^\mu_{-\frac{1}{2}+it}(\cosh a)}{Q^\mu_{c-\frac{1}{2}}(\cosh a)} \left\{ (i\tau - c) \, \cosh a \, P^\mu_{-\frac{1}{2}+i\tau}(\cosh a) \, Q^\mu_{c-\frac{1}{2}}(\cosh a) \right. \right. \\
& \left. + \left(\mu + c - \frac{1}{2}\right) P^\mu_{-\frac{1}{2}+i\tau}(\cosh a) \, Q^\mu_{c-\frac{1}{2}}(\cosh a) \right. \\
& \left. - \left(\mu + i\tau - \frac{1}{2}\right) P^\mu_{-\frac{3}{2}+i\tau}(\cosh a) \, Q^\mu_{c-\frac{1}{2}}(\cosh a) \right\} \\
& + (\tau^2 + c^2)(t^2 - \tau^2)^{-1} \left\{ i(\tau - t) \, \cosh a \, P^\mu_{-\frac{1}{2}+i\tau}(\cosh a) \right. \\
& \times P^\mu_{-\frac{1}{2}+it}(\cosh a) + \left(\mu + it - \frac{1}{2}\right) P^\mu_{-\frac{1}{2}+i\tau}(\cosh a) \, P^\mu_{-\frac{3}{2}+it}(\cosh a) \\
& \left. \left. - \left(\mu + i\tau - \frac{1}{2}\right) P^\mu_{-\frac{3}{2}+i\tau}(\cosh a) \, P^\mu_{-\frac{3}{2}+it}(\cosh a) \right\} \right]
\end{aligned}
$$

128

and

$$B(\tau) = \frac{\mu_1 \left\{ \Gamma(\frac{1}{2} - \mu + i\tau)\, \Gamma(\frac{1}{2} - \mu - i\tau) \right\}^2 \sinh^2 \pi\tau}{\pi \left\{ \lambda_1\, \tau\, \Gamma(\frac{1}{2} - \mu + i\tau)\, \Gamma(\frac{1}{2} - \mu - i\tau)\, \sinh \pi\tau + \mu_1\, (\tau^2 + c^2) \right\}}$$

$$\times \left[\int_0^a (\tau^2 + c^2)\, g(x)\, P^\mu_{-\frac{1}{2}+i\tau}(\cosh x)\, \sinh x\, dx + \frac{g(a)}{Q^\mu_{c-\frac{1}{2}}(\cosh a)} \right.$$

$$\times \left\{ (i\tau - c)\, \cosh a\, P^\mu_{-\frac{1}{2}+i\tau}(\cosh a)\, Q^\mu_{c-\frac{1}{2}}(\cosh a) \right.$$

$$+ \left(\mu + c - \frac{1}{2} \right) P^\mu_{-\frac{1}{2}+i\tau}(\cosh a)\, Q^\mu_{c-\frac{3}{2}}(\cosh a)$$

$$\left. \left. - \left(\mu + i\tau - \frac{1}{2} \right) P^\mu_{-\frac{3}{2}+i\tau}(\cosh a)\, Q^\mu_{c-\frac{1}{2}}(\cosh a) \right\} \right].$$

It is easy to verify that by putting $\mu = 0$ in the dual integral equations discussed above, some results obtained in Section 3.1 for the solutions of dual integral equations involving Legendre function as kernel are recovered.

3.3 Kernels involving generalized associated Legendre functions

In this section we consider certain dual integral equations involving generalized associated Legendre functions of the first kind (cf. Mandal 1995). Solutions of these are obtained by using properties of the generalized associated Legendre functions and the inversion formula for the generalized associated Legendre functions of the first kind (cf. Mandal and Mandal 1997, p. 57). These dual integral equations are more general dual integral equations than those which are presented in Sections 3.1 and 3.2. Some of these are also considered in the book of Virchenko (1989), p. 86-89.

We consider the dual integral equations

$$\int_0^\infty f(\tau)\, P^{\mu,\nu}_{-\frac{1}{2}+i\tau}(\cosh x)\, d\tau = g(x), \; 0 \le x \le a, \tag{3.183}$$

$$\int_0^\infty m(\tau)\, f(\tau)\, P^{\mu,\nu}_{-\frac{1}{2}+i\tau}(\cosh x)\, d\tau = h(x), \; a < x < \infty, \tag{3.184}$$

where $m(\tau) = \tau \sinh 2\pi\tau \; \Gamma(\frac{1-\mu+\nu}{2} + i\tau) \; \Gamma(\frac{1-\mu+\nu}{2} - i\tau) \; \Gamma(\frac{1-\mu-\nu}{2} + i\tau) \; \Gamma(\frac{1-\mu-\nu}{2} - i\tau)$

and $P^{\mu,\nu}_{-\frac{1}{2}+i\tau}(\cosh x)$ is the generalized associated Legendre function of the first kind.

To obtain the solution of this pair, multiplying equation (3.183) by

$$\pi^{-1} \, 2^{(\mu-\nu)/2} \, \Gamma\left(\frac{1}{2} - \mu\right) \, \cos\pi\mu \, \sinh^{1-\mu} x \, (\cosh t - \cosh x)^{\mu-\frac{1}{2}} \, (\cosh t + 1)^{(\nu-\mu)/2}$$

$$\times \, (\cosh x + 1)^{(\mu-\nu)/2} \, F\left(\frac{\mu-\nu}{2}, \frac{1+\mu-\nu}{2}; \frac{1}{2}+\mu; \frac{\cosh t - \cosh x}{1 + \cosh t}\right), \qquad (3.185)$$

integrating with respect to x from 0 to t, and then differentiating with respect to t and interchanging the order of integration and using the relation (A.3.5), we find that

$$F_c\left[f(\tau)\right] = g_1(t), \; 0 \le t \le a, \qquad (3.186)$$

where

$$g_1(t) = \pi^{-1} \, 2^{(\mu-\nu)/2} \, \Gamma\left(\frac{1}{2} - \mu\right) \, \cos\pi\mu \, \frac{d}{dt}\left\{(\cosh t + 1)^{(\nu-\mu)/2}\right.$$

$$\times \int_0^t \frac{(\cosh x + 1)^{(\mu-\nu)/2}}{(\cosh t - \cosh x)^{\frac{1}{2}-\mu}} \, F\left(\frac{\mu-\nu}{2}, \frac{1+\mu-\nu}{2}; \frac{1}{2}+\mu; \frac{\cosh t - \cosh x}{1 + \cosh t}\right)$$

$$\left. \times \sinh^{1-\mu} x \; g(x) \; dx\right\} \qquad (3.187)$$

Similarly, multiplying the equation (3.184) by

$$\pi^{-2} \, 2^{((\mu-\nu)/2)-1} \left\{\Gamma\left(\frac{1}{2} - \mu\right)\right\}^{-1} (\cosh x - \cosh t)^{-\mu-\frac{1}{2}}$$

$$\times \, F\left(\frac{-\mu+\nu}{2}, -\frac{\nu+\mu}{2}; \frac{1}{2} - \mu; \frac{\cosh x - \cosh t}{1 + \cosh x}\right) \sinh^{1+\mu} x, \qquad (3.188)$$

integrating with respect to x from t to ∞, and then interchanging the order of integration and using the relation (A.3.10), we get

$$F_c\left[f(\tau)\right] = h_1(t), \; a < t < \infty, \qquad (3.189)$$

where

$$h_1(t) = \pi^{-2} \, 2^{((\mu-\nu)/2)-1} \left\{\Gamma\left(\frac{1}{2} - \mu\right)\right\}^{-1} \int_t^\infty (\cosh x - \cosh t)^{-\mu-\frac{1}{2}}$$

$$\times F\left(\frac{\nu-\mu}{2}, -\frac{\nu+\mu}{2}; \frac{1}{2} - \mu; \frac{\cosh x - \cosh t}{1 + \cosh x}\right) \sinh^{1+\mu} x$$

$$\times h(x) \; dx. \qquad (3.190)$$

130

Hence, by the use of the Fourier cosine inversion theorem, equations (3.186) and (3.189) give

$$f(\tau) = \sqrt{\frac{2}{\pi}} \left[\int_0^a g_1(t) \, \cos \tau t \, dt + \int_a^\infty h_1(t) \, \cos \tau t \, dt \right], \qquad (3.191)$$

which gives the solution of the pair (3.183) and (3.184).

Next, we consider the pair

$$\int_0^\infty \tau \, f(\tau) \, P^{\mu,\nu}_{-\frac{1}{2}+i\tau}(\cosh x) \, d\tau \;=\; p(x), \; 0 \le x \le a, \qquad (3.192)$$

$$\int_0^\infty \tau^{-1} \, m(\tau) \, f(\tau) \, P^{\mu,\nu}_{-\frac{1}{2}+i\tau}(\cosh x) \, d\tau \;=\; q(x), \; x > a. \qquad (3.193)$$

Multiplying equation (3.192) by the expression (3.185) and integrating with respect to x from 0 to t, then interchanging the order of integration and using the relation (A.3.12), we obtain

$$F_s \left[f(\tau) \right] = p_1(t), \; 0 \le t \le a, \qquad (3.194)$$

where

$$
\begin{aligned}
p_1(t) \;=\;& \pi^{-1} \, 2^{(\mu-\nu)/2} \, \Gamma\left(\frac{1}{2} - \mu\right) \, \cos \pi\mu \, (\cosh t + 1)^{(\nu-\mu)/2} \\
& \times \int_0^t (\cosh t - \cosh x)^{\mu - \frac{1}{2}} (\cosh x + 1)^{(\mu-\nu)/2} \\
& \times F\left(\frac{\mu - \nu}{2}, \frac{1 + \mu - \nu}{2}; \frac{1}{2} + \mu; \frac{\cosh t - \cosh x}{1 + \cosh t}\right) \, \sinh^{1-\mu} x \\
& \times p(x) \, dx.
\end{aligned}
\qquad (3.195)
$$

Similarly, multiplying equation (3.193) by the expression (3.188) and integrating with respect to x from t to ∞, and then differentiating with respect to t and interchanging the order of integrations and using the relation (A.3.11), we get

$$F_s \left[f(\tau) \right] = q_1(t), \; a < t < \infty, \qquad (3.196)$$

where

$$
\begin{aligned}
q_1(t) \;=\;& -\pi^{-1} \, 2^{((\mu-\nu)/2)-1} \left\{ \Gamma\left(\frac{1}{2} - \mu\right) \right\}^{-1} \frac{d}{dt} \int_t^\infty (\cosh x - \cosh t)^{-\mu - \frac{1}{2}} \\
& \times F\left(\frac{\nu - \mu}{2}, -\frac{\nu + \mu}{2}; \frac{1}{2} - \mu; \frac{\cosh x - \cosh t}{1 + \cosh x}\right) \, \sinh^{1+\mu} x \\
& \times q(x) \, dx.
\end{aligned}
\qquad (3.197)
$$

131

Therefore, using the inversion formula for the Fourier sine transform, the relations (3.194) and (3.196) produce

$$f(\tau) = \sqrt{\frac{2}{\pi}} \left[\int_0^a p_1(t) \, \sin \tau t \, dt + \int_a^\infty q_1(t) \, \sin \tau t \, dt \right], \qquad (3.198)$$

which gives the solution of the dual integral equations (3.192) and (3.193).

Now, we consider the more general dual integral equations

$$\int_0^\infty [1 + \omega(\tau)] \, f(\tau) \, P_{-\frac{1}{2}+i\tau}^{\mu,\nu} (\cosh x) \, d\tau = g(x), \ 0 \le x \le a, \qquad (3.199)$$

$$\int_0^\infty m(\tau) \, f(\tau) \, P_{-\frac{1}{2}+i\tau}^{\mu,\nu} (\cosh x) \, d\tau = h(x), \ x > a, \qquad (3.200)$$

where $\omega(\tau)$ is a known weight function and $m(\tau)$ is the same as in equation (3.184).

Equation (3.199) can be written as

$$\int_0^\infty f(\tau) \, P_{-\frac{1}{2}+i\tau}^{\mu,\nu} (\cosh x) \, d\tau = g(x) - \int_0^\infty \omega(\tau) \, f(\tau) \, P_{-\frac{1}{2}+i\tau}^{\mu,\nu} (\cosh x) \, d\tau, \ 0 \le x \le a. \qquad (3.201)$$

Then by the relation (3.191), the solution of this pair is obtained as

$$f(\tau) = \sqrt{\frac{2}{\pi}} \left[\int_0^a G(t) \, \cos \tau t \, dt + \int_a^\infty h_1(t) \, \cos \tau t \, dt \right], \qquad (3.202)$$

where $h_1(t)$ is the same as defined by relation (3.190), but

$$
\begin{aligned}
G(t) \ = \ & \pi^{-1} \, 2^{(\mu-\nu)/2} \, \Gamma\left(\frac{1}{2} - \mu\right) \cos \, \pi\mu \, \frac{d}{dt} \left[(\cosh t + 1)^{(\nu-\mu)/2} \right. \\
& \times \int_0^t (\cosh t - \cosh x)^{\mu-\frac{1}{2}} (\cosh x + 1)^{(\mu-\nu)/2} \\
& \times F\left(\frac{\mu-\nu}{2}, \frac{1+\mu-\nu}{2}; \frac{1}{2} + \mu; \frac{\cosh t - \cosh x}{1 + \cosh t} \right) \, \sinh^{1-\mu} x \\
& \times \left\{ g(x) - \int_0^\infty \omega(\tau) f(\tau) \, P_{-\frac{1}{2}+i\tau}^{\mu,\nu} (\cosh x) d\tau \right\} dx \left. \right], \ 0 \le t \le a. (3.203)
\end{aligned}
$$

Hence, the solution of the pair (3.199) and (3.200) can be reduced to the solution of a Fredholm integral equation of the second kind for the function $G(t)$ which can be written in the form

$$G(t) + \int_0^a K(t, u) \, G(u) \, du = G^*(t), \ 0 \le t \le a, \qquad (3.204)$$

where

$$K(t, u) = (2\pi)^{-\frac{1}{2}} \{\omega_c(t + u) - \omega_c(|t - u|)\}$$

$$\omega_c(t) = F_c[\omega(\tau)],$$

$$G^*(t) = g_1(t) - g_2(t),$$

where $g_1(t)$ is defined by the relation (3.187),

$$g_2(t) = F_c[\omega(\tau) H_1(\tau)],$$

with

$$H_1(\tau) = \sqrt{\frac{2}{\pi}} \int_a^\infty h_1(t) \cos \tau t \, dt.$$

Similarly, the solution of the general pair of integral equations

$$\int_0^\infty \tau f(\tau) P^{\mu,\nu}_{-\frac{1}{2}+i\tau} (\cosh x) \, d\tau = p(x), \ 0 \le x \le a,$$

$$\int_0^\infty \tau^{-1} [1 + \omega(\tau)] m(\tau) f(\tau) P^{\mu,\nu}_{-\frac{1}{2}+i\tau} (\cosh x) \, d\tau = q(x), \ x > a,$$

is given by

$$f(\tau) = \sqrt{\frac{2}{\pi}} \left[\int_0^a p_1(t) \sin \tau t \, dt + \int_a^\infty Q(t) \sin \tau t \, dt \right],$$

where $p_1(t)$ is the same as defined by the relation (3.195), but $Q(t)$ satisfies the Fredholm integral equation of the second kind given by

$$Q(t) = Q^*(t) + \int_a^\infty K_1(t, u) Q(u) \, du, \ a < t < \infty,$$

where

$$K_1(t, u) = (2\pi)^{-\frac{1}{2}} \{\omega_c(t + u) - \omega_c(|t - u|)\},$$

$$Q^*(t) = q_1(t) - q_2(t),$$

$q_1(t)$ is defind by the relation (3.197), and

$$q_2(t) = \sqrt{\frac{2}{\pi}} \int_0^a p_1(y) \left\{ \int_0^\infty \omega(\tau) \sin \tau t \sin \tau y \, d\tau \right\} dy.$$

Chapter 4

Dual integral equations with trigonometric function kernel

Dual integral equations with trigonometric functions as kernel arise frequently in the analysis of some mixed boundary value problems in a plane, such as crack problems in the two-dimensional theory of elasticity (cf. Sneddon 1966; Sneddon and Lowengrub 1969; Virchenko 1989, p. 71). In this chapter, we present some new methods of solution of a number of dual integral equations involving trigonometric functions as kernel which are of recent interest. These arise in some other types of physical problems also, such as problems of the linearised theory of water waves for which two applications are illustrated. We also consider here solutions of certain dual integral equations by using generalized Mehler−Fock and generalized associated Mehler−Fock integral transform theorems.

4.1 Some elementary methods

4.1.1 In this section, we first consider a simple pair of integral equations, studied by Nasim and Aggarwala (1984), of the following type:

$$\int_0^\infty f(t) \, \cos xt \, dt \; = \; g(x), \; 0 < x < 1, \tag{4.1}$$

$$\int_0^\infty f(t) \, \sin xt \, dt \; = \; h(x), \; 1 < x < \infty. \tag{4.2}$$

This pair arises in the crack problems in the classical theory of elasticity. The solution of this pair is obtained by considering the following two cases.

(I) We consider the pair

$$\int_0^\infty f(t)\ \cos xt\ dt\ =\ g(x),\ 0 < x < 1, \tag{4.3}$$

$$\int_0^\infty f(t)\ \sin xt\ dt\ =\ 0,\ 1 < x < \infty. \tag{4.4}$$

Equation (4.4) can be written as

$$\sqrt{\frac{2}{\pi}} \int_0^\infty f(t)\ \sin xt\ dt = H(1-x)\phi(x),\ 0 < x < \infty, \tag{4.5}$$

for some unknown function $\phi(x)$, $H(x)$ being the Heaviside unit function. By the Fourier sine inversion theorem, the relation (4.5) gives

$$f(t) = \sqrt{\frac{2}{\pi}} \int_0^1 \phi(x)\ \sin xt\ dx. \tag{4.6}$$

Using the integral representation (cf. Erdélyi et al. (1954b), p. 7)

$$\sin xt = t \int_0^x \frac{u\ J_0(ut)}{\sqrt{x^2 - u^2}}\ du,$$

equation (4.6) becomes

$$\begin{aligned}
f(t) &= \sqrt{\frac{2}{\pi}}\ t \int_0^1 \phi(x) \left\{ \int_0^1 \frac{u\ J_0(xt)}{\sqrt{x^2-u^2}}\ du \right\} dx \\
&= \sqrt{\frac{2}{\pi}}\ t \int_0^1 u\ J_0(ut) \left\{ \int_u^1 \frac{\phi(x)}{\sqrt{x^2-u^2}}\ dx \right\} du \\
&= t \int_0^1 \psi(u)\ J_0(ut)\ du,
\end{aligned} \tag{4.7}$$

where

$$\psi(u) = \sqrt{\frac{2}{\pi}}\ u \int_0^1 \frac{\phi(x)}{\sqrt{x^2-u^2}}\ dx. \tag{4.8}$$

Now, equation (4.3) can be written as

$$g(x) = \frac{d}{dx} \int_0^\infty t^{-1}\ f(t)\ \sin xt\ dt,\ 0 < x < 1. \tag{4.9}$$

Substituting the value of $f(t)$ from (4.7) into (4.9), we get

$$\begin{aligned}
g(x) &= \frac{d}{dx} \int_0^\infty \sin xt \left\{ \int_0^1 \psi(u)\ J_0(ut)\ du \right\} dt \\
&= \frac{d}{dx} \int_0^1 \psi(u) \left\{ \int_0^\infty \sin xt\ J_0(ut)\ dt \right\} du.
\end{aligned} \tag{4.10}$$

135

Using the result (A.1.2), equation (4.10) reduces to an inverse Abel-type integral equation

$$g(x) = \frac{d}{dx} \int_0^x \frac{\psi(u)}{\sqrt{x^2 - u^2}} \, du, \ 0 < x < 1, \tag{4.11}$$

whose solution is

$$\psi(u) = \frac{2u}{\pi} \int_0^u \frac{g(x)}{\sqrt{u^2 - x^2}} \, dx. \tag{4.12}$$

Therefore, from the relations (4.7) and (4.12), we obtain

$$f(t) = \frac{2t}{\pi} \int_0^1 u \, J_0(ut) \left\{ \int_0^u \frac{g(x) \, dx}{\sqrt{u^2 - x^2}} \right\} \, du, \tag{4.13}$$

which gives the solution of the pair (4.3) and (4.4).

(II) Now, we consider the pair

$$\int_0^\infty f(t) \, \cos xt \, dt \ = \ 0, \ 0 < x < 1, \tag{4.14}$$

$$\int_0^\infty f(t) \, \sin xt \, dt \ = \ h(x), \ 1 < x < \infty. \tag{4.15}$$

From (4.14), using the Fourier cosine inversion theorem, we get

$$f(t) = \sqrt{\frac{2}{\pi}} \int_1^\infty \psi(x) \, \cos xt \, dx, \tag{4.16}$$

for an appropriately defined unknown function $\psi(x)$. Now, using the result (A.1.5), equation (4.16) reduces, after changing the order of integration, to

$$f(t) = t \int_1^\infty \chi(u) \, J_0(ut) \, du, \tag{4.17}$$

where

$$\chi(u) = \sqrt{\frac{2}{\pi}} \int_1^u \frac{\psi(x)}{\sqrt{u^2 - x^2}} \, dx.$$

Now equation (4.15) is equivalent to

$$h(x) = -\frac{d}{dx} \int_0^\infty t^{-1} f(t) \, \cos xt \, dt, \ 1 < x < \infty. \tag{4.18}$$

Substituting the value of $f(t)$ from (4.17) into (4.18), we get

$$h(x) = -\frac{d}{dx} \int_0^\infty \cos xt \left\{ \int_1^\infty \chi(u) \, J_0(ut) \, du \right\} dt$$

$$= -\frac{d}{dx} \int_1^\infty \chi(u) \left\{ \int_0^\infty \cos xt \, J_0(ut) \, dt \right\} du$$

$$= -\frac{d}{dx} \int_x^\infty \frac{\chi(u)}{\sqrt{u^2 - x^2}} \, du, \quad 1 < x < \infty.$$

The solution of the above integral equation is given by

$$\chi(u) = \frac{2u}{\pi} \int_u^\infty \frac{h(x) \, dx}{\sqrt{x^2 - u^2}}.$$

Hence, from (4.17), we obtain

$$f(t) = \frac{2t}{\pi} \int_1^\infty u \, J_0(ut) \left\{ \int_u^\infty \frac{h(x) \, dx}{\sqrt{x^2 - u^2}} \right\} du. \tag{4.19}$$

This gives the solution of the pair (4.14) and (4.15). Now combining the relations (4.13) and (4.19), we get

$$f(t) = \frac{2t}{\pi} \int_0^1 u \, J_0(ut) \left\{ \int_0^u \frac{g(x) \, dx}{\sqrt{u^2 - x^2}} \right\} du$$

$$+ \frac{2t}{\pi} \int_1^\infty u \, J_0(ut) \left\{ \int_u^\infty \frac{h(x) \, dx}{\sqrt{x^2 - u^2}} \right\} du, \tag{4.20}$$

which is the solution of the dual integral equations (4.1) and (4.2).

The solution of various types of dual integral equations with trigonometric kernels can be deduced from the relation (4.20). We mention here one particular case as given below.

The equation (4.3) can be written as

$$\int_0^\infty t^{-1} f(t) \, \sin xt \, dt = \int_0^x g(x) \, dx, \quad 0 < x < 1,$$

$$= G(x), \quad \text{say}, \quad 0 < x < 1.$$

Then the dual integral equations (4.3) and (4.4) become

$$\int_0^\infty t^{-1} f(t) \, \sin xt \, dt = G(x), \quad 0 < x < 1,$$

$$\int_0^\infty f(t) \, \sin xt \, dt = 0, \quad 1 < x < \infty.$$

The solution of this pair of integral equations is then derived from the relation (4.20) to give

$$f(t) = \frac{2t}{\pi} \int_0^1 u \, J_0(ut) \left\{ \int_0^u \frac{G'(x) \, dx}{\sqrt{u^2 - x^2}} \right\} du.$$

If we put $G(x) = x$, then

$$f(t) = J_1(x)$$

which is the solution of the pair

$$\int_0^\infty t^{-1} f(t) \, \sin xt \, dt = x, \ 0 < x < 1,$$

$$\int_0^\infty f(t) \, \sin xt \, dt = 0, \ 1 < x < \infty.$$

4.1.2 Here, we have analysed two different sets of dual integral equations with trigonometric function kernels for their solutions by exploiting the behaviour of one of the integrals of the dual integral equations at the *turning point* (cf. Chakrabarti and Mandal 1998). This method has already been introduced in Section 2.1 of Chapter 2 for dual integral equations with first-kind Bessel function kernel. We now consider dual integral equations with cosine and sine kernels separately.

Case I: Dual integral equations involving cosine function kernel

(A) We first consider the pair

$$\int_0^\infty \frac{f(t)}{t} \, \cos xt \, dt = g(x), \ 0 < x < 1, \tag{4.21}$$

$$\int_0^\infty f(t) \, \cos xt \, dt = 0, \ 1 < x < \infty, \tag{4.22}$$

where $g(x)$ is known. To obtain the solution of this pair, keeping in mind the square root singularity behaviour of the integral in (4.22) as $x \to 1-$, which is dictated by the physics of the mixed boundary value problem in which this pair of integral equations arises, we set

$$\int_0^\infty f(t) \, \cos xt \, dt = \frac{d}{dx} \left[x \int_x^1 \frac{\psi(s) \, ds}{\sqrt{s^2 - x^2}} \right], \ 0 < x < 1, \tag{4.23}$$

138

where $\psi(s)$ is an unknown differentiable function to be determined such that $\psi(1) \neq 0$.

The transformation $s = xu$ in the integral followed by differentiation applied to the right side of (4.23) shows that it can be expressed as

$$-\frac{\psi(1)}{\sqrt{1-x^2}} + \int_x^1 \frac{\{\psi(s) + s\,\psi'(s)\}}{\sqrt{s^2 - x^2}}\,ds,$$

and this exhibits the singularity structure clearly as $x \to 1 - 0$.

Using the Fourier cosine inversion theorem and the result (A.1.9), from the relations (4.22) and (4.23), we obtain

$$f(t) = t \int_0^1 s\,J_1(st)\,\psi(s)\,ds. \tag{4.24}$$

Substituting $f(t)$ from (4.24) into (4.21) and by using the result (A.1.3) along with the Abel inversion formula (cf. Sneddon 1966, p. 41), the unknown function $\psi(s)$ is obtained in the form

$$\psi(s) = -\frac{2}{\pi}\frac{d}{ds}\int_0^s \frac{g(u)\,du}{\sqrt{s^2 - u^2}}, \quad 0 < s < 1. \tag{4.25}$$

Therefore, the equations (4.24) and (4.25) produce the solution of the pair (4.21) and (4.22) in the form

$$\begin{aligned}
f(t) &= \frac{2t^2}{\pi}\int_0^1 s\,J_0(st)\left\{\int_0^s \frac{g(u)\,du}{\sqrt{s^2 - u^2}}\right\}ds \\
&\quad - \frac{2t\,J_1(t)}{\pi}\int_0^1 \frac{g(u)\,du}{\sqrt{1 - u^2}}.
\end{aligned} \tag{4.26}$$

This solution completely agrees with the one obtained earlier by Busbridge (1938) who used the Mellin transform technique to solve dual integral equations.

(B) The second pair for our consideration is

$$\int_0^\infty f(t)\,\cos xt\,dt = g(x), \quad 0 < x < 1, \tag{4.27}$$

$$\int_0^\infty \frac{f(t)}{t}\,\cos xt\,dt = 0, \quad 1 < x < \infty. \tag{4.28}$$

139

The solution of this pair can be obtained by setting

$$\int_0^\infty \frac{f(t)}{t} \cos xt \, dt = \int_x^1 \frac{\psi(s) \, ds}{\sqrt{s^2 - x^2}}, \quad 0 < x < 1, \tag{4.29}$$

where $\psi(s)$ is a new unknown function to be determined. We emphasize here that the substitution (4.29) takes care of the required behaviour of the integral on the left of the relation (4.28) at the turning point $x = 1$, where it is expected to be $O(\sqrt{1-x})$, as $x \to 1-$. This behaviour arises due to the physics of the corresponding mixed boundary value problem in which these integral equations arise.

Utilizing the Fourier cosine inversion theorem and the result (A.1.8), equations (4.28) and (4.29) give

$$f(t) = t \int_0^1 J_0(st) \, \psi(s) \, ds. \tag{4.30}$$

Substituting this value of $f(t)$ from the relation (4.30) into (4.27) and then using the result (A.1.1) along with the Abel inversion formula, we obtain

$$\psi(s) = \frac{2s}{\pi} \int_0^s \frac{g(u) \, du}{\sqrt{s^2 - u^2}}, \quad 0 < s < 1. \tag{4.31}$$

Thus equations (4.30) and (4.31) give the solution of the pair (4.27) and (4.28) as

$$f(t) = \frac{2t}{\pi} \int_0^1 s \, J_0(st) \left\{ \int_0^s \frac{g(u) \, du}{\sqrt{s^2 - u^2}} \right\} ds. \tag{4.32}$$

As a special case, (4.32) agrees with Titchmarsh (1937, p. 339).

Case II: Dual integral equations involving sine function kernel

(A) First, we analyse the pair

$$\int_0^\infty \frac{f(t)}{t} \sin xt \, dt = g(x), \quad 0 < x < 1, \tag{4.33}$$

$$\int_0^\infty f(t) \sin xt \, dt = 0, \quad 1 < x < \infty. \tag{4.34}$$

To obtain the solution of this pair, we set

$$\int_0^\infty f(t) \sin xt \, dt = \frac{d}{dx} \int_x^1 \frac{\psi(s) \, ds}{\sqrt{s^2 - x^2}}, \quad 0 < x < 1, \tag{4.35}$$

140

where $\psi(s)$ is another unknown differentiable function to be determined, such that $\psi(1) \neq 0$.

Using the Fourier sine inversion theorem and the result (A.1.8), from the relations (4.34) and (4.35), we obtain

$$f(t) = -t \int_0^1 J_0(st) \; \psi(s) \; ds. \tag{4.36}$$

Now, using the result (A.1.2) and Abel's inversion formula, from the relations (4.33) and (4.36), the unknown function can be obtained as

$$\begin{aligned}
\psi(s) &= -\frac{2}{\pi} \frac{d}{ds} \int_0^s \frac{u \; g(u)}{\sqrt{s^2 - u^2}} \; du \\
&= -\frac{2s}{\pi} \int_0^s \frac{g'(u) \; du}{\sqrt{s^2 - u^2}}, \quad 0 < s < 1, \tag{4.37}
\end{aligned}$$

where $g(0) = 0$ has been used (which is apparent if we make $x \to 0+$ on both sides of (4.33)). Therefore, from the relations (4.36) and (4.37), the solution of the pair (4.33) and (4.34) is given by

$$f(t) = \frac{2t}{\pi} \int_0^1 s \; J_0(st) \left\{ \int_0^s \frac{g'(u) \; du}{\sqrt{s^2 - u^2}} \right\} ds. \tag{4.38}$$

This agrees with the solution of one of the special cases of the dual integral equations handled earlier by Busbridge (1938).

(B) Secondly, we consider the pair

$$\int_0^\infty f(t) \; \sin xt \; dt = g(x), \; 0 < x < 1, \tag{4.39}$$

$$\int_0^\infty \frac{f(t)}{t} \sin xt \; dt = 0, \; 1 < x < \infty. \tag{4.40}$$

The solution of this pair can be obtained by setting

$$\int_0^\infty \frac{f(t)}{t} \sin xt \; dt = x \int_x^1 \frac{\psi(s)}{\sqrt{s^2 - x^2}} \; ds, \; 0 < x < 1, \tag{4.41}$$

where $\psi(s)$ is an unknown differentiable function to be determined.

Using the Fourier sine inversion theorem and the result (A.1.9), equations (4.40) and (4.41) give rise to the following expression for $f(t)$:

$$f(t) = t \int_0^1 s \; J_1(st) \; \psi(s) \; ds. \tag{4.42}$$

141

Integrating the relation (4.42) in equation (4.39), we find

$$\frac{d}{dx} \int_0^1 s \, \psi(s) \left\{ \int_0^\infty J_1(st) \, \cos xt \, dt \right\} ds = -g(x), \; 0 < x < 1. \qquad (4.43)$$

Now, using the result (A.1.3) and integrating the equation (4.43) on both sides, we find that

$$\int_0^x \frac{\psi(s) \, ds}{\sqrt{x^2 - s^2}} = \frac{1}{x} \int_0^x g(u) \, du, \; 0 < x < 1. \qquad (4.44)$$

This is an Abel integral equation, whose solution is

$$\psi(s) = -\frac{2}{\pi s} \int_0^s \frac{u \, g(u) \, du}{\sqrt{s^2 - u^2}}, \; 0 < s < 1. \qquad (4.45)$$

Thus, the solution of the pair (4.39) and (4.40) can be obtained from (4.42) and (4.45), and we get

$$f(t) = \frac{2t}{\pi} \int_0^1 J_1(st) \left\{ \int_0^s \frac{u \, g(u)}{\sqrt{s^2 - u^2}} \, du \right\} ds. \qquad (4.46)$$

This agrees with one of the special cases of the solution obtained by Noble (1963) earlier.

4.1.3 Recently, Aggarwala and Nasim (1996) considered two pairs of dual integral equations whose kernels involve a combination of both sine and cosine functions. These pairs of integral equations arise in the solution of mixed boundary value problems of steady state temperatures in a quarter plane describe below.

We consider a quarter plane whose edge $x = 0$ is losing heat to the environment at zero temperature according to Newton's law of cooling while on the edge $y = 0$, the temperature is controlled on a portion of this edge and the heat input is known on the remaining part. This problem is formulated mathematically as follows. We have to find a function $v(x, y)$ $(x > 0, \; y > 0)$ which satisfies

$$\frac{\partial^2 v}{\partial x^2} + \frac{\partial^2 v}{\partial y^2} = 0, \quad x > 0, \; y > 0, \qquad (4.47)$$

$$\frac{\partial v}{\partial x} - \alpha v = 0, \qquad x = 0, \qquad (4.48)$$

142

where α is a non-zero real constant, and either

$$(i) \quad v(x,0) = g(x), 0 < x < 1, \tag{4.49}$$

$$\frac{\partial v}{\partial y}(x,0) = -h(x), x > 1, \tag{4.50}$$

or

$$(ii) \quad \frac{\partial v}{\partial y}(x,0) = -g(x), 0 < x < 1, \tag{4.51}$$

$$v(x,0) = h(x), \ x > 1. \tag{4.52}$$

Also in each case $|v|$ must be bounded at infinity.

An appropriate representation for $v = v(x,y)$ is

$$v(x,y) = \int_0^\infty (\alpha \ \sin xt + t \ \cos xt) e^{-ty} \ f(t) \ dt, \ x > 0, \ y > 0, \tag{4.53}$$

where $f(t)$ is an unknown function to be determined. Using the above two kinds of mixed boundary conditions, the problem is reduced to one of the following two pairs of integral equations:

$$\int_0^\infty (\alpha \ \sin xt + t \ \cos xt) \ f(t) \ dt = g(x), \ 0 < x < 1, \tag{4.54}$$

$$\int_0^\infty t(\alpha \ \sin xt + t \ \cos xt) \ f(t) \ dt = h(x), \ x > 1, \tag{4.55}$$

and

$$\int_0^\infty t(\alpha \ \sin xt + t \ \cos xt) \ f(t) \ dt = g(x), \ 0 < x < 1, \tag{4.56}$$

$$\int_0^\infty (\alpha \ \sin xt + t \ \cos xt) \ f(t) \ dt = h(x), \ x > 1. \tag{4.57}$$

To obtain the solutions of the above pairs, we assume that the integrals
$\int_0^\infty f(t) \ \sin xt \ dt, \int_0^\infty t \ f(t) \ \sin xt \ dt, \int_0^\infty t \ f(t) \ \cos xt \ dt$ and $\int_0^\infty t^2 \ f(t) \ \cos xt \ dt$
exist. Then, we set

$$F(x) = \int_0^\infty f(t) \ \sin xt \ dt \ \text{ so that } F'(x) = \int_0^\infty t \ f(t) \ \cos xt \ dt, \tag{4.58}$$

for $0 < x < 1$, and

$$G(x) = \int_0^\infty t \ f(t) \ \sin xt \ dt \ \text{ so that } G'(x) = \int_0^\infty t^2 \ f(t) \ \cos xt \ dt, \tag{4.59}$$

143

for $1 < x < \infty$, where the dash denotes differentiation with respect to x.

Equation (4.58) implies that $F(0+) = F(0) = 0$. Then the pair (4.54) and (4.55) reduce to

$$\alpha \, F(x) + F'(x) = g(x), \ 0 < x < 1, \tag{4.60}$$

$$\alpha \, G(x) + G'(x) = h(x), \ x > 1, \tag{4.61}$$

with the condition that

$$F(0) = 0. \tag{4.62}$$

For the pair (4.56) and (4.57), we set

$$F(x) = \int_0^\infty t \, f(t) \, \sin xt \, dt, \ 0 < x < 1, \tag{4.63}$$

and

$$G(x) = \int_0^\infty f(t) \, \sin xt \, dt, \ x > 1, \tag{4.64}$$

so that we again get equations (4.60) and (4.61) with the condition (4.62).

For both cases, equations (4.60) and (4.61) produce

$$F(x) = e^{-\alpha x} \int_0^x e^{\alpha t} \, g(t) \, dt, \ 0 < x < 1, \tag{4.65}$$

and

$$G(x) = e^{-\alpha x} \int_1^x e^{\alpha t} \, h(t) \, dt + Ce^{-\alpha x}, \ x > 1, \tag{4.66}$$

where C is an unknown constant to be determined by the condition that $v(x,0)$ is continuous at $x = 1$.

(I) Solution of the pair (4.54) and (4.55).

This pair is reduced to the pair of integral equations

$$\int_0^\infty f(t) \, \sin xt \, dt \ = \ F(x) = e^{-\alpha x} \int_0^x e^{\alpha t} \, g(t) \, dt, \ 0 < x < 1, \tag{4.67}$$

$$\int_0^\infty t \, f(t) \, \sin xt \, dt \ = \ G(x) = e^{-\alpha x} \int_1^x e^{\alpha t} \, h(t) \, dt + Ce^{-\alpha x}, \ x > 1. \tag{4.68}$$

Following the method presented in §4.1.1, the solution of this pair is obtained as

$$f(t) = \int_0^1 u\, J_0(ut)\, g_1(u)\, du + \int_1^\infty u\, J_0(ut)\, h_1(u)\, du$$

$$+ \frac{2C}{\pi} \int_1^\infty u\, J_0(ut) \left\{ \int_u^\infty \frac{e^{-\alpha x} dx}{\sqrt{x^2 - u^2}} \right\} du, \tag{4.69}$$

where

$$g_1(u) = \frac{2}{\pi} \frac{d}{du} \int_0^u \frac{x\, F(x)}{\sqrt{u^2 - x^2}}\, dx = \frac{2}{\pi} \int_0^u \frac{F'(x)\, dx}{\sqrt{u^2 - x^2}}, \tag{4.70}$$

$$h_1(u) = \frac{2}{\pi} \int_u^\infty \frac{e^{-\alpha x}}{\sqrt{x^2 - u^2}} \left\{ \int_1^\infty e^{\alpha t}\, h(t)\, dt \right\} dx. \tag{4.71}$$

We have used here the condition $F(0) = 0$ in deriving the relation (4.70). To find the constant C, we use the continuity condition

$$\lim_{x \to 1+} v(x, 0) = \lim_{x \to 1-} v(x, 0) = \alpha\, F(1) + F'(1) = g(1). \tag{4.72}$$

Using the result

$$\int_u^\infty \frac{e^{-\alpha x}}{\sqrt{x^2 - u^2}}\, dx = K_0(\alpha u),$$

where K denotes the modified Bessel function, we have

$$\lim_{x \to 1+} v(x, 0) = \lim_{x \to 1+} \left(\alpha\, U(x) + U'(x) \right),$$

where for $x > 1$

$$U(x) = \int_0^\infty f(t)\, \sin xt\, dt = \int_0^1 \frac{u\, g_1(u)}{\sqrt{x^2 - u^2}}\, du + \int_1^x \frac{u\, h_1(u)}{\sqrt{x^2 - u^2}}\, du$$

$$+ \frac{2C}{\pi} \int_1^x \frac{u\, K_0(\alpha u)}{\sqrt{x^2 - u^2}}\, du. \tag{4.73}$$

By integration by parts, (4.73) becomes

$$U(x) = \left[h_1(1) - g_1(1) + \frac{2C}{\pi} K_0(\alpha) \right] \sqrt{x^2 - 1} + \int_0^1 g_1'(u)\, \sqrt{x^2 - u^2}\, du$$

$$+ \int_1^x h_1'(u)\, \sqrt{x^2 - u^2}\, du - \frac{2C}{\pi} \int_1^x \alpha\, K_1(\alpha u)\, \sqrt{x^2 - u^2}\, dx$$

$$+ x\, g_1(0). \tag{4.74}$$

We observe in this relation that the function $U'(x)$ will be bounded as $x \to 1+$ if the coefficient of $\sqrt{x^2 - 1}$ is zero. Hence,

$$C = \frac{\pi}{2} \frac{g_1(1) - h_1(1)}{K_0(\alpha)}. \tag{4.75}$$

This value of C gives

$$
\begin{aligned}
\alpha\, U(x) + U'(x) &= \alpha \int_0^1 g_1'(u)\, \sqrt{x^2 - u^2}\, du + \alpha \int_1^x h_1'(u)\, \sqrt{x^2 - u^2}\, du \\
&\quad + \int_0^1 \frac{x\, g_1'(u)}{\sqrt{x^2 - u^2}}\, du + \int_1^x \frac{x\, h_1'(u)}{\sqrt{x^2 - u^2}}\, du \\
&\quad - \frac{2C}{\pi} \int_1^x \alpha^2\, K_1(\alpha u)\, \sqrt{x^2 - u^2}\, du - \frac{2C}{\pi} \int_1^x \frac{\alpha x\, K_1(\alpha u)}{\sqrt{x^2 - u^2}}\, du \\
&\quad + (1 + \alpha x)\, g_1(0). \tag{4.76}
\end{aligned}
$$

Hence, we have

$$\lim_{x \to 1+} \left(\alpha\, U(x) + U'(x) \right) = \alpha \int_0^1 \frac{u\, g_1(u)}{\sqrt{1 - u^2}}\, du + \int_0^1 \frac{g_1'(u)}{\sqrt{1 - u^2}}\, du + g_1(0).$$

Now,

$$
\begin{aligned}
g_1(u) &= \frac{2}{\pi} \int_0^u \frac{F'(x)\, dx}{\sqrt{u^2 - x^2}} \\
&= \frac{2u}{\pi} F''(0) + \frac{2}{\pi} \int_0^u \left(\frac{F'(x) - F'(0)}{x} \right)' \sqrt{u^2 - x^2}\, dx + F'(0). \tag{4.77}
\end{aligned}
$$

Hence,

$$g_1'(u) = \frac{2}{\pi} F''(0) + \frac{2}{\pi} \int_0^u \left(\frac{F'(x) - F'(0)}{x} \right)' \frac{u}{\sqrt{u^2 - x^2}}\, dx. \tag{4.78}$$

Using the relations (4.77) and (4.78), interchanging the order of integration and using the result

$$\int_x^y \frac{u\, du}{\sqrt{(u^2 - x^2)(y^2 - u^2)}} = \frac{\pi}{2}, \quad y > x > 0, \tag{4.79}$$

we obtain that

$$
\begin{aligned}
\lim_{x \to 1+} \left(\alpha\, U(x) + U'(x) \right) &= \alpha\, (F(1) - F(0)) + F''(0) \\
&\quad + (F'(1) - F'(0)) - F''(0) + g_1(0) \\
&= \alpha\, F(1) + F'(1). \tag{4.80}
\end{aligned}
$$

146

Thus, the relations (4.72) and (4.80) show that $v(x, 0)$ is continuous at $x = 1$. It can also be shown that if C is given by relation (4.75), then under suitable restrictions on the data, $v(x, 0)$ as given by (4.76), is bounded as $x \to \infty$.

(II) Solution of the pair (4.56) and (4.57).

In this case, the dual integral equations (4.56) and (4.59) are reduced to

$$\int_0^\infty t\, f(t)\, \sin xt\, dt = F(x) = e^{-\alpha x} \int_0^x g(t)\, e^{-\alpha t}\, dt,\ 0 < x < 1, \tag{4.81}$$

$$
\begin{aligned}
\int_0^\infty f(t)\, \sin xt\, dt &= e^{-\alpha x} \int_1^x e^{\alpha t}\, h(t)\, dt + De^{-\alpha x},\ x > 1 \\
&= p(x) + De^{-\alpha x},\ \text{say.}
\end{aligned}
\tag{4.82}
$$

The solution of the above dual integral equations is now given by

$$
\begin{aligned}
f(t) &= \frac{2}{\pi} \int_0^1 J_1(ut)\, m(u)\, du - \frac{2}{\pi} \int_1^\infty u\, J_1(ut)\, n(u)\, du \\
&\quad + \frac{2D\alpha}{\pi} \int_1^\infty u\, J_1(ut)\, K_1(\alpha u)\, du,
\end{aligned}
\tag{4.83}
$$

where

$$m(u) = \int_0^u \frac{x\, F(x)}{\sqrt{u^2 - x^2}}\, dx, \tag{4.84}$$

$$n(u) = \frac{d}{du} \int_u^\infty \frac{p(x)\, dx}{\sqrt{x^2 - u^2}}. \tag{4.85}$$

Proceeding as in Case I and assuming that the function $h(x)$ is restricted so that $n(x) = 0$ for $x < 1$, then we get

$$v(x, 0) = \alpha\, U(x) + U'(x),\ 0 < x < 1,$$

where

$$
\begin{aligned}
U(x) &= \frac{2x}{\pi} \sqrt{1 - x^2}\, (m(1) + n(1) - D\alpha\, K_1(\alpha)) \\
&\quad + \frac{2x}{\pi} \int_1^\infty \left[\frac{n(u) - D\alpha\, K_1(\alpha u)}{u} \right]' \sqrt{u^2 - x^2}\, du \\
&\quad - \frac{2x}{\pi} \int_x^1 \left(\frac{m(u)}{u^2} \right)' \sqrt{u^2 - x^2}\, du.
\end{aligned}
\tag{4.86}
$$

147

For $v(x, 0)$ to be continuous at $x = 1$, we must have

$$D = \frac{m(1) + n(1)}{\alpha\, K_1(\alpha)}. \tag{4.87}$$

This value of D gives

$$\lim_{x \to 1-} \left(\alpha\, U(x) + U'(x) \right) = \frac{2(1 + \alpha)}{\pi} \int_1^\infty \left(\frac{n(u)}{u} \right)' \sqrt{u^2 - 1}\; du$$
$$- \frac{2}{\pi} \int_1^\infty \left(\frac{n(u)}{u} \right)' \frac{du}{\sqrt{u^2 - 1}}. \tag{4.88}$$

Also,

$$n(u) = \frac{d}{du} \int_u^\infty \frac{p(x)\, dx}{\sqrt{x^2 - u^2}}$$

gives

$$p(x) = -\frac{2x}{\pi} \int_x^\infty \frac{n(u)\, du}{\sqrt{u^2 - x^2}}. \tag{4.89}$$

Differentiating both sides of equation (4.89) and then substituting it into the relation (4.88), we find that

$$\lim_{x \to 1-} \left(\alpha\, U(x) + U'(x) \right) = \alpha\, p(1) + p'(1) = h(1).$$

This shows that $v(x, 0)$ is continuous at $x = 1$ when D is given by the relation (4.87).

The above two pairs of integral equations are studied when α is a non-zero real constant. Now for $\alpha = 0$, the integral representation (4.53) is not valid. The correct representation for $v(x, y)$ is

$$v(x, y) = C_1 + \int_0^\infty t\, f(t)\, \cos xt\; e^{-ty}\, dt, \tag{4.90}$$

where C_1 is an arbitrary constant. Then the mixed boundary value problem is reduced to the following two pairs of dual integral equations.

(i) Find C_1 and $f(t)$ such that

$$C_1 + \int_0^\infty t\, f(t)\, \cos xt\, dt = g(x),\ 0 < x < 1, \tag{4.91}$$

$$\int_0^\infty t^2\, f(t)\, \cos xt\, dt = h(x),\ x > 1; \tag{4.92}$$

148

and

(ii)

$$\int_0^\infty t^2\, f(t)\, \cos xt\, dt \;=\; g(x),\; 0 < x < 1, \tag{4.93}$$

$$C_1 + \int_0^\infty t\, f(t)\, \cos xt\, dt \;=\; h(x),\; x > 1. \tag{4.94}$$

In case (ii), the constant C_1 is to be determined so that $|h(x) - C_1| \to 0$ as $x \to \infty$. In case (i), the constant C_1 is to be determined such that $v(x,0)$ is continuous at $x = 1$.

The pair (4.91) and (4.92) can be written as

$$\int_0^\infty t\, f(t)\, \cos xt\, dt \;=\; g(x) - C_1,\; 0 < x < 1, \tag{4.95}$$

$$\int_0^\infty t\, f(t)\, \sin xt\, dt \;=\; -\int_x^\infty h(x)dx = h_1(x),\; \text{say},\; x > 1. \tag{4.96}$$

Then, as before, we obtain

$$\begin{aligned}
f(t) \;=\;& \frac{2}{\pi}\int_0^1 u\, J_0(ut)\, G_1(u)\, du + \frac{2}{\pi}\int_1^\infty u\, J_0(ut)\, H_1(u)\, du \\
& -C_1 \int_0^1 u\, J_0(ut)\, du,
\end{aligned} \tag{4.97}$$

where

$$G_1(u) = \int_0^u \frac{g(x)\, dx}{\sqrt{u^2 - x^2}} \quad \text{and} \quad H_1(u) = \int_u^\infty \frac{h_1(x)\, dx}{\sqrt{x^2 - u^2}}. \tag{4.98}$$

For $x > 1$ we have

$$v(x,0) - C_1 = \frac{d}{dx}\int_0^\infty f(t)\, \sin xt\, dt = U'(x),\; \text{say}. \tag{4.99}$$

Substituting $f(t)$ from (4.97) into (4.99) and then simplifying, we obtain

$$\begin{aligned}
U(x) \;=\;& \int_0^\infty f(t)\, \sin xt\, dt \\
\;=\;& \frac{2}{\pi}\sqrt{x^2 - 1}\left[H_1(1) - G_1(1) + \frac{\pi C_1}{2}\right] \\
& + \frac{2x}{\pi}G_1(0) - C_1 x + \frac{2}{\pi}\int_0^1 G_1'(u)\, \sqrt{x^2 - u^2}\, du \\
& + \frac{2}{\pi}\int_1^x H_1'(u)\, \sqrt{x^2 - u^2}\, du.
\end{aligned} \tag{4.100}$$

149

Since $v(x, 0)$ is to be continuous at $x = 1$, we must have

$$C_1 = \frac{2}{\pi}\left[H_1(1) - G_1(1)\right]. \tag{4.101}$$

Using the above value of C_1, we observe that

$$
\begin{aligned}
\lim_{x \to 1+} v(x, 0) &= \lim_{x \to 1+} U'(x) + C_1 \\
&= \frac{2}{\pi} G_1(0) + \frac{2}{\pi} \int_0^1 \frac{G_1'(u)}{\sqrt{1 - u^2}}\, du.
\end{aligned} \tag{4.102}
$$

From the relation (4.98), we get

$$G_1(u) = \int_0^u \frac{g(x)\, dx}{\sqrt{u^2 - x^2}}$$

so that

$$g(x) = \frac{2}{\pi} \frac{d}{dx} \int_0^x \frac{u\, G_1(u)}{\sqrt{x^2 - u^2}}\, du.$$

$$= \frac{2}{\pi} G_1(0) + \frac{2x}{\pi} \int_0^x \frac{G_1'(u)}{\sqrt{x^2 - u^2}}\, du.$$

Therefore, the relation (4.102) becomes

$$\lim_{x \to 1+} v(x, 0) = g(1),$$

which shows that $v(x, 0)$ is continuous at $x = 1$. It can also be seen from (4.100) that if $h(x)$ is suitably restricted then $v(x, 0)$ is bounded as $x \to \infty$.

For case (ii), the solution is given by the relation (4.83) in the limit $\alpha \to 0+$.

Some special cases of the dual integral equations

We consider the pair

$$\int_0^\infty f(t)\, (\alpha \sin xt + t \cos xt)\, dt = g(x), \ 0 < x < 1, \tag{4.103}$$

$$\int_0^\infty f(t)\, (\alpha \sin xt + t \cos xt)\, dt = 0, \ x > 1, \tag{4.104}$$

with the additional requirement that the quantity

$$\int_0^\infty f(t)\, (\alpha\, \sin xt + t\, \cos xt)\, dt$$

150

is continuous at $x = 1$.

(1) $g(x) = 1$, $0 < x < 1$. Then

$$f(t) = \int_0^1 u\, J_0(ut)\, F(u)\, du + \frac{F(1)}{K_0(\alpha)} \int_1^\infty u\, J_0(ut)\, K_0(\alpha u)\, du,$$

where

$$F(u) = \frac{2}{\pi} \int_0^u \frac{e^{-\alpha x}}{\sqrt{u^2 - x^2}}\, dx.$$

(2) $g(x) = 1 + \alpha x$, $0 < x < 1$. Then

$$f(t) = \frac{J_1(t)}{t} + \frac{1}{K_0(\alpha)} \int_1^\infty u\, J_0(ut)\, K_0(\alpha u)\, du.$$

(3) $g(x) = \alpha x^2 + 2x$, $0 < x < 1$. Then

$$f(t) = \frac{4}{\pi} \int_0^1 u^2\, J_0(ut)\, du + \frac{4}{\pi K_0(\alpha)} \int_1^\infty u\, J_0(ut)\, K_0(\alpha u)\, du.$$

4.1.4. Eswaran (1990) showed that there exists a class of two-dimensional diffraction problems which can be reduced to a standard form of dual integral equations. These problems involve diffraction of electromagnetic waves by a finite strip, a finite slit, the diffraction of scalar or vector elastic waves by a rigid strip or crack, etc. A general method for solving such dual integral equations is given by Eswaran (1990) by the artifice of constructing a set of functions of compact support biorthogonal to another given set of functions. Here we consider the following simple pair of dual integral equations to illustrate the method (cf. Eswaran 1990):

$$\int_{-\infty}^\infty \sqrt{t^2 - k^2}\, f(t)\, e^{ixt} dt \;=\; g(x), \ |x| < 1, \tag{4.105}$$

$$\int_{-\infty}^\infty f(t)\, e^{ixt} dt \;=\; 0, \ |x| > 1, \tag{4.106}$$

where k is a given real constant. This pair arises in problems of the diffraction of H-polarized light by a finite strip of infinite conductivity, and also in problems of the diffraction of sound waves by a finite rigid barrier or that of a shear waves by a finite crack. The solution of the above pair determines an integral representation of the scattered field completely, which is given by

$$\phi^s(x, y) = \int_{-\infty}^\infty f(t)\, \exp\left[ixt - \sqrt{t^2 - k^2}\, |y| \right] dt. \tag{4.107}$$

151

It can be proved that the above representation (4.107) is always valid for a function $f(t)$ whose Fourier transform has compact support (cf. Eswaran 1990). Then the function $f(t)$ can be expanded in the form of a Neumann series as

$$f(t) = \sum_{n=1}^{\infty} b_n \, \frac{J_n(t)}{t}, \tag{4.108}$$

where the b_n's are constants. For such an expression of the function $f(t)$ given by (4.108), equation (4.106) is automatically satisfied, since for $n > 0$,

$$\int_{-\infty}^{\infty} \frac{J_n(t)}{t} \, e^{ixt} dt = \begin{cases} -\dfrac{2 \, i^{in+1}}{n} \sqrt{1 - x^2} \, U_{n-1}(x), & |x| < 1, \\ 0, & |x| > 1, \end{cases} \tag{4.109}$$

$U_n(x)$ being the Chebyshev polynomial of the second kind.

We define a function

$$\psi_m(x) = (m + 1) \, i^{m+2} \int_{-\infty}^{\infty} \sqrt{t^2 - k^2} \, \frac{J_{m+1}(t)}{t} \, e^{ixt} \, dt, \tag{4.110}$$

where the constants $(m + 1)i^{m+2}$ have been introduced for convenience.

Now, we represent the solution of the pair (4.105) and (4.106) as

$$f(t) = \sum_{m=0}^{\infty} (m + 1) \, i^{m+2} \, a_m \, \frac{J_{m+1}(t)}{t}, \tag{4.111}$$

where the a_m's are unknown constants to be determined. Substituting (4.111) into (4.105) and then using the relation (4.110), we find that

$$\sum_{m=0}^{\infty} a_m \, \psi_m(x) = g(x), \ |x| < 1. \tag{4.112}$$

Equation (4.106) is satisfied identically for the representation (4.111). Now we construct some functions $\phi_n(x)$ which satisfy the conditions

$$\phi_n(x) = 0, \ |x| > 1; \ n = 0, 1, 2, \ldots \tag{4.113}$$

and

$$\int_{-1}^{1} \psi_m(x) \, \phi_n(x) \, dx = \delta_{mn}. \tag{4.114}$$

This set of functions $\phi_n(x)$ is thus biorthogonal to the set of functions $\psi_m(x)$ and the functions $\phi_n(x)$ are functions of compact support. Then, multiplying equation (4.112) by $\phi_n(x)$ and integrating term by term, we obtain that

$$a_n = \int_{-1}^{1} g(x) \, \phi_n(x) \, dx. \tag{4.115}$$

We can also assume that the function $\psi_m(x)$ can be expanded in terms of Chebyshev polynomials for the region $|x| < 1$ as

$$\psi_m(x) = \sum_{k=0}^{\infty} C_k^m \, U_k(x). \tag{4.116}$$

From the representation (4.110) and using the result (4.109), we can take the product of (4.116) with the conjugate of (4.109), integrate for all x and then using the relations

$$\int_{-1}^{1} \sqrt{1 - x^2} \, U_m(x) \, U_n(x) \, dx = \frac{\pi}{2} \, \delta_{mn}, \tag{4.117}$$

and

$$\frac{1}{2\pi} \int_{-\infty}^{\infty} e^{i(u-u')x} dx = \delta(u - u'), \tag{4.118}$$

we find

$$C_n^m = \frac{2(m+1)(n+1)}{(-i)^{m+1}(-i)^{n+1}} \int_{-\infty}^{\infty} \frac{\sqrt{t^2 - k^2}}{t^2} \, J_{m+1}(t) \, J_{n+1}(t) \, dt. \tag{4.119}$$

It is seen from the symmetry of the integrand that C_n^m will vanish unless m and n are both even or both odd. Hence, if m and n are both even or both odd, we need to integrate (4.119) only from 0 to ∞ and double the value, and also note that $C_n^m = 0$ if m is even and n is odd or vice versa. To construct the functions $\phi_n(x)$ with the help of C_n^m, we write

$$\phi_n(x) = \begin{cases} \sum_{r=0}^{\infty} D_r^n \, U_r(x) \, \sqrt{1 - x^2}, & |x| < 1, \\ 0, & |x| > 1. \end{cases} \tag{4.120}$$

Now, since we require the relation (4.114) to hold, we find from (4.116) and (4.120) that

$$\sum_{k=0}^{\infty} C_k^m \, D_k^n = \delta_{mn}. \tag{4.121}$$

153

We denote the infinite dimensional matrix

$$[C] = [C_i^j] \quad \text{and} \quad [D] = [D_i^j]$$

where C_i^j and D_i^j represent the ij-th elements of $[C]$ and $[D]$ respectively. Then, the relation (4.121) is equivalent to

$$[D] = [C^T]^{-1}.$$

Since $[C]$ is a symmetric matrix,

$$[D] = [C]^{-1}.$$

Hence, we observe that the inverse of the matrix $[C]$ gives the matrix $[D]$. Therefore, the coefficients D_k^n and thus from (4.120), the functions $\phi_n(x)$, are determined uniquely.

A number of two-dimensional diffraction problems involving scalar and vector waves can be reduced to dual integral equations. The solution of the dual integral equations can be obtained by the method mentioned above. For applications of the above method to some physical problems of diffraction, one can see the original paper of Eswaran (1990).

4.1.5 Applications to water wave problems

(A) Rolling of a plate

As applications to a physical problem of dual integral equations, we consider two-dimensional motion in water due to small rolling oscillations of a thin rigid vertical plate partially immersed in deep water up to a depth a below the mean free surface (cf. Mandal, Banerjea and Kanoria 1997). This represents a model of a rolling ship in the classical linearised theory of water waves. The water occupies the region $y \geq 0$ with $y = 0$ as the mean free surface. The plate is hinged at the point $(0, b)$ ($b < a$) and performs small rolling oscillations with amplitude θ_0 and frequency σ.

Assuming linear theory and irrotational motion, the motion in water can be described by a real-valued potential function $\Phi(x, y, t)$, which is the real part of $\phi(x, y)e^{-i\sigma t}$ where the complex-valued function $\phi(x, y)$ satisfies the following boundary value problem:

$$\nabla^2 \phi = 0, \quad \text{for } y \geq 0, \tag{4.122}$$

with

$$K\phi + \phi_y = 0, \quad \text{on } y = 0, \tag{4.123}$$

where $K = \frac{\sigma^2}{g}$, g being the acceleration due to gravity.

$$\phi_x = i\sigma\theta_0 (y - b), \quad \text{on } x = 0, \text{ for } 0 < y < a, \tag{4.124}$$

$$r^{\frac{1}{2}}\nabla\phi \text{ is bounded as } r = \{x^2 + (y - a)^2\}^{\frac{1}{2}} \to 0, \tag{4.125}$$

$$\phi, \nabla\phi \to 0 \text{ as } y \to \infty, \tag{4.126}$$

and

$$\phi(x, y) \to \begin{cases} A_0 \ e^{-Ky+iKx}, & \text{as } x \to \infty, \\ B_0 \ e^{-Ky-iKx}, & \text{as } x \to -\infty, \end{cases} \tag{4.127}$$

where A_0, B_0 denote the unknown (complex) amplitudes of the wave motion set up by the rolling oscillations of the plate at large distances on its right and left sides. Utilizing Havelock's expansion of the water wave potential, the solution of the partial differential equation (4.122) satisfying the conditions (4.123), (4.126) and (4.127) can be expressed in the form

$$\phi(x, y) = \begin{cases} A_0 \ e^{-Ky+iKx} + \int_0^\infty A(k) \ L(k, y) \ e^{-kx} \ dk, & x > 0, \\ B_0 \ e^{-Ky-iKx} + \int_0^\infty B(k) \ L(k, y) \ e^{kx} \ dk, & x < 0, \end{cases} \tag{4.128}$$

with $L(k, y) = k \ \cos ky - K \sin ky$, where $A(k)$ and $B(k)$ are unknown functions.

Since $\phi_x(0+, y) = \phi_x(0-, y)$, for $y \geq 0$, by using Havelock's inversion theorem (cf. Ursell (1947)), we find that

$$A_0 = -B_0, \ A(k) = -B(k). \tag{4.129}$$

155

Now using the representation (4.128), the condition (4.124) produces

$$iKA_0 \, e^{-Ky} - \int_0^\infty k \, A(k) \, L(k,y) \, dk = i\sigma\theta_0(y-b), \ 0 < y < a. \tag{4.130}$$

The condition of continuity of $\phi(x,y)$ across the gap below the plate along with the utilization of the relation (4.129) gives

$$A_0 e^{-Ky} + \int_0^\infty A(k) \, L(k,y) \, dk = 0, \ a < y < \infty. \tag{4.131}$$

Equations (4.131) and (4.132) can be simplified by applying the operator $\left(\frac{d}{dy} + K\right)$ to both sides to obtain the integral equations for the determination of the function $A(k)$. These are given by

$$\int_0^\infty A(k) \, k(k^2 + K^2) \, \sin ky \, dk \ = \ i\sigma\theta_0\{K(y-b)+1\}, \ 0 < y < a, \tag{4.132}$$

$$\int_0^\infty A(k) \, (k^2 + K^2) \, \sin ky \, dk \ = \ 0, \ a < y < \infty. \tag{4.133}$$

Substituting

$$f(k) = (k^2 + K^2) \, A(k), \tag{4.134}$$

the dual integral equations (4.132) and (4.133) become

$$\frac{d}{dy} \int_0^\infty \frac{f(k)}{k} \, \sin ky \, dk \ = \ iS(y) + C, \ 0 < y < a, \tag{4.135}$$

$$\int_0^\infty f(k) \, \sin ky \, dk \ = \ 0, \ a < y < \infty, \tag{4.136}$$

where

$$S(y) = \sigma\theta_0 \left\{ \frac{K^2 y^2}{2} + (1 - Kb)y \right\}, \tag{4.137}$$

and C is an arbitrary constant.

The above pair of integral equations are also equivalent to the pair (4.3) and (4.4). Here we use a somewhat different method to solve this pair. For this, we set

$$h(y) = \int_0^\infty f(k) \, \sin ky \, dk, \ 0 < y < a. \tag{4.138}$$

By the Fourier inversion theorem, the equations (4.138) and (4.137) produce

$$f(k) = \frac{2}{\pi} \int_0^a h(u) \, \sin ku \, du. \tag{4.139}$$

156

Let us denote

$$g(y) = \phi(0+, y) - \phi(0-, y), \ 0 < y < a. \tag{4.140}$$

Then relations (4.128) and (4.129) give

$$A_0 e^{-Ky} + \int_0^\infty A(k) \, L(k, y) \, dk = \frac{1}{2} \, g(y), \ 0 < y < a. \tag{4.141}$$

Applying the operator $(\frac{d}{dy} + K)$ on both sides of (4.141), we find that

$$\int_0^\infty f(k) \, \sin ky \, dk = -\frac{1}{2} \left\{ g'(y) + Kg(y) \right\}, \ 0 < y < a. \tag{4.142}$$

Thus, from (4.138) and (4.142), we get

$$h(y) = -\frac{1}{2} \left\{ g'(y) + Kg(y) \right\}. \tag{4.143}$$

From the condition (4.125) and the relation (4.143), we must have

$$h(y) = O \left(|a - y|^{-\frac{1}{2}} \right) \ \text{ as } y \to a. \tag{4.144}$$

This suggests that we can represent $h(y)$ as

$$h(y) = \frac{d}{dy} \left[\int_y^a \frac{t \, p(t)}{\sqrt{t^2 - y^2}} \, dt \right], \ 0 < y < a, \tag{4.145}$$

where $p(t)$ is an unknown but bounded function in $(0, a)$ and $p(a) \neq 0$.
From equations (4.139) and (4.145), we obtain

$$f(k) = \frac{2}{\pi} \int_0^a \sin ku \, \frac{d}{du} \left\{ \int_u^a \frac{t \, p(t)}{\sqrt{t^2 - u^2}} \, dt \right\} du. \tag{4.146}$$

Using this $f(k)$, equation (4.136) reduces to an Abel type integral equation of the first kind as

$$\int_0^y \frac{t \, p(t)}{\sqrt{y^2 - t^2}} \, dt = i\sigma\theta_0 \left(\frac{Ky^3}{6} + \frac{1 - Kb}{2} \, y^2 \right) + Cy, \ 0 < y < a. \tag{4.147}$$

The solution of this integral equation is

$$p(t) = i\sigma\theta_0 \left\{ \frac{Kt^2}{4} - \frac{2}{\pi} (Kb - 1)t \right\} + C, \ 0 < t < a. \tag{4.148}$$

157

Substituting the expression for $p(t)$ from (4.148) into (4.146), after simplification, we obtain

$$f(k) = -\frac{2}{\pi} \int_0^a \frac{u \, \sin ku}{\sqrt{a^2 - u^2}} \left[\sigma\theta_0 \left\{ \frac{2i}{\pi}(Kb - 1) \left(-a + \sqrt{a^2 - u^2} \, ln \, \frac{a + \sqrt{a^2 - u^2}}{u} \right) \right. \right.$$
$$\left. \left. -\frac{iK}{4}(a^2 - 2u^2) + C \right\} \right] du. \tag{4.149}$$

The relations (4.130), (4.134) and (4.149) produce one equation for the determination of the unknown constants A_0 and C which is given by

$$A_0 = \frac{2}{\pi} \beta + M + iCa \, K_1(Ka), \tag{4.150}$$

with

$$\beta = \sigma\theta_0 \left[-\pi a^2 K_2(Ka) - \left\{ \frac{\pi}{8} Ka + (1 - Kb) \right\} a^2 K_1(Ka) \right.$$
$$+ \frac{\pi a}{2K}(1 - Kb) \left\{ K_0(Ka) \, L_1(Ka) + L_0(Ka) K_1(Ka) \right\}$$
$$\left. + \frac{\pi^2}{4K^2} e^{-Ka} \left\{ (Ka + 2)^2 - 2Kb(Ka + 1) \right\} \right], \tag{4.151}$$

$$M = \frac{\sigma\theta_0}{K^2} \left[2(1 - Kb) - \frac{1}{2} e^{-Ka} \left\{ (Ka + 2)^2 - 2Kb(Ka + 1) \right\} \right], \tag{4.152}$$

where $K_n(x)$ and $L_n(x)$ $(n = 0, 1)$ are the second-kind n-th order modified Bessel function and Struve function respectively.

In a similar way, the relation (4.131) produces another equation for A_0 and C given by

$$A_0 = -\frac{2}{\pi} iN - C\pi a \, I_1(Ka), \tag{4.153}$$

where

$$N = \sigma\theta_0\pi \left[-\frac{\pi}{4} a^2 I_2(Ka) + \left\{ \frac{\pi}{8} Ka + (1 - Kb) \right\} a^2 I_1(Ka) \right.$$
$$\left. + \frac{\pi a}{2K}(1 - Kb) \left\{ I_0(Ka)L_1(Ka) - L_0(Ka)I_1(Ka) \right\} \right], \tag{4.154}$$

$I_n(x), (n = 0, 1)$ being the first-kind modified Bessel function.

From equations (4.150) and (4.153), the constants C and A_0 are obtained as

$$C = -\frac{1}{\Delta a} \left[\frac{2}{\pi}(iN + \beta) + M \right], \tag{4.155}$$

$$A_0 = -\frac{\sigma\theta_0 \, \pi a}{\Delta K} \left[\frac{1}{2} - \frac{(1 - Kb)}{Ka} \left\{ L_1(Ka) + I_1(Ka) \right\} \right], \tag{4.156}$$

where

$$\Delta = \pi\, I_1(Ka) + iK_1(Ka). \tag{4.157}$$

Therefore, the solution of the dual integral equations is now obtained as

$$
\begin{aligned}
A(k) &= \frac{\sigma\theta_0 a}{k^2 + K^2}\left[(U + iV)a\, J_1(ka) - \frac{iKa}{2k}J_2(ka)\right.\\
&\quad\left. +\frac{i(1 - Kb)}{K}\left\{J_1(ka)\, H_0(Ka) - H_1(Ka)\, J_0(ka)\right\}\right],
\end{aligned}
\tag{4.158}
$$

$$U = \frac{\pi}{\Delta_1 Ka}\left[\frac{1}{2} + \frac{(Kb - 1)}{Ka}\left\{I_1(Ka) + L_1(Ka)\right\}\right], \tag{4.159}$$

$$
\begin{aligned}
V &= \frac{1}{I_1(Ka)}\left[\frac{1}{2}I_2(Ka) + \frac{(1 - Kb)}{Ka}\left\{I_1(Ka)\, L_0(Ka) - L_1(Ka)\, I_0(ka)\right\}\right.\\
&\quad\left. -\frac{U}{\pi}\, K_1(Ka)\right],
\end{aligned}
\tag{4.160}
$$

and

$$\Delta_1 = K_1^2(Ka) + \pi^2 I_1^2(Ka).$$

The explicit expressions of $\phi(x, y)$ for $x > 0$ and $x < 0$ are now obvious. Hence, the real-valued potential function $\Phi(x, y, t)$ describing the motion is obtained.

(B) Wave scattering by thin vertical barrier

Chakrabarti et al. (1995, 1997) considered three basic problems of scattering of surface water waves in the linearised set-up involving (i) a fully submerged thin vertical barrier, (ii) a partially immersed thin vertical barrier and (iii) a fully submerged thin vertical plate. Mathematically speaking, the three problems, referred to as P_1, P_2 and P_3 respectively, are the following.

Determine three harmonic functions ϕ_1, ϕ_2 and ϕ_3 of two variables x and y, representing rectangular cartesian coordinates of a point in two-dimensions with $y > 0$, in the forms:

$$
\phi_j(x, y) = \begin{cases}
T_j\, e^{-Ky + iKx} + \displaystyle\int_0^\infty A_j(k)\, L(k, y)\, e^{-kx}\, dk, & x > 0 \\[2mm]
e^{-Ky + iKx} + R_j\, e^{-Ky - iKx} + \displaystyle\int_0^\infty B_j(k)\, L(k, y)\, e^{kx}\, dk, & x < 0,
\end{cases}
\tag{4.161}
$$

with

$$L(k, y) = k\, \cos ky - K\, \sin ky, \tag{4.162}$$

159

where $A_j(k), B_j(k)$ $(j = 1, 2, 3)$ are unknown functions, and T_j and R_j are unknown constants to be determined by utilizing the following conditions:

$$(a) \ \frac{\partial \phi_j}{\partial x} \ \text{is continuous on} \ x = 0, \ \text{for all} \ y,$$

$$(b) \ \frac{\partial \phi_j}{\partial x} = 0 \ \text{on} \ x = 0\pm, \ \text{for} \ y \ \in L_j, \qquad\qquad (4.163)$$

$$(c) \ \phi_j \ \text{is continuous on} \ x = 0, \ \text{for} \ y \ \in G_j,$$

where L_j represents the interval $a_j < y < b_j$ and $G_j = (0, \infty) - L_j$, with $a_1 = a, b_1 = \infty$ (corresponding to P_1), $a_2 = 0, b_2 = b$ (corresponding to P_2) and $a_3 = c, b_3 = d \ (c > 0, d > 0, d > c)$ (corresponding to P_3). The derivatives $\frac{\partial \phi_j}{\partial x}(0, y)$ will have square root singularities at the turning points $y = a$ for $P_1, y = b$ for P_2, $y = c, d$ for P_3. It may be noted that G_3 is a double interval. Using the condition (a) along with Havelock's expansion theorem (cf. Ursell 1947), we find that

$$A_j(k) = -B_j(k) \qquad\qquad (4.164)$$

and

$$T_j + R_j = 1 \ (j = 1, 2, 3). \qquad\qquad (4.165)$$

Then the conditions (b) and (c) above give rise to the following integral equations (dual for $j = 1, 2$ and triple for $j = 3$)

$$
\begin{aligned}
\int_0^\infty A_j(k) \ L(k, y) \ dk &= R_j \ e^{-Ky}, \ y \ \in G_j, \\
\int_0^\infty k A_j(k) \ L(k, y) \ dk &= iK(1 - R_j) \ e^{-Ky}, \ y \ \in L_j.
\end{aligned}
\qquad (4.166)
$$

Keeping in mind the singular behaviour of the integrals on the left side of the second equation in (4.166) at the turning points, we integrate these relations with respect to y and recast them into the form

$$\int_0^\infty A_j(k) \ (k \sin ky + K \ \cos ky) \ dk = -i(1 - R_j) \ e^{-Ky} - D_j, \ y \ \in L_j, \qquad (4.167)$$

where the D_j's are arbitrary constants.

160

Equations (4.167) can also be represented as

$$\frac{d}{dy} \int_0^\infty \frac{A_j(k)}{k} L(k,y) \, dk = i(1 - R_j) \, e^{-Ky} + D_j, \; y \in L_j. \tag{4.168}$$

Thus, by using the representation (4.168), the integral equations (4.166) of our concern take the following forms:

$$\int_0^\infty A_j(k) \, L(k,y) \, dk \quad = \quad R_j \, e^{-Ky}, \; y \in G_j,$$

$$\frac{d}{dy} \int_0^\infty \frac{A_j(k)}{k} L(k,y) \, dk \quad = \quad i(1 - R_j) \, e^{-Ky} + D_j, \; y \in L_j. \tag{4.169}$$

These equations can be further expressed, after using the operator $(\frac{d}{dy} + K)$ on both sides, formally, in the forms:

$$\int_0^\infty F_j(k) \, \sin ky \, dk \quad = \quad 0, \; y \in G_j,$$

$$\frac{d}{dy} \int_0^\infty \frac{F_j(k)}{k} \sin ky \, dk \quad = \quad C_j, \; y \in L_j, \tag{4.170}$$

with

$$F_j(k) = (k^2 + K^2) \, A_j(k), \; (j = 1, 2, 3) \tag{4.171}$$

where the C_j's are arbitrary constants.

We also note that, because of the Riemann–Lebesgue lemma, we must have $C_1 = 0$ for problem P_1. The constants C_2 and C_3 associated with the problems P_2 and P_3 respectively remain arbitrary and we shall determine them fully during the course of the mathematical analysis.

Setting

$$h_j(y) = \int_0^\infty F_j(k) \, \sin ky \, dk, \; y \in (0, \infty), \; (j = 1, 2, 3), \tag{4.172}$$

and using the Fourier sine inversion formula on (4.172) and the first of the relations (4.170), we obtain that

$$F_j(k) = \frac{2}{\pi} \int_{L_j} h_j(y) \, \sin ky \, dy. \tag{4.173}$$

161

Then, substituting (4.173) in the second equation of (4.170) and using the well-known result

$$\int_0^\infty \frac{\sin kt \; \sin ky}{k} \; dk = ln \left| \frac{y+t}{y-t} \right|,$$

we derive the following singular integral equations for the determination of the functions $h_j(y)$:

$$\int_{L_j} \frac{2t \; h_j(t)}{y^2 - t^2} \; dt = \frac{\pi}{2} C_j, \; y \; \in L_j, \; (j = 1, 2, 3) \tag{4.174}$$

where the integral is in the sense of the Cauchy principal value.

In order to solve the singular integral equations (4.174), we must know the behaviour of the functions $h_j(t)$ at the end points of L_j. For this, we define $f_j(y)$ as

$$f_j(y) = \phi_j(0+, y) - \phi_j(0-, y), \; (j = 1, 2, 3). \tag{4.175}$$

Then using (4.161), in conjunction with (4.164) and (4.165), we find that

$$\int_0^\infty A_j(k) \; L(k, y) \; dk = \frac{1}{2} \; f_j(y) + R_j \; e^{-Ky}, \; y \; \in L_j, \tag{4.176}$$

while

$$f_j(y) = 0 \; \text{ for } y \; \in G_j, \tag{4.177}$$

because of the continuity of ϕ_j across G_j.

Applying the operator $\frac{d}{dy} + K$ to both sides of (4.176), we then find

$$\int_0^\infty F_j(k) \; \sin ky \; dk = \frac{1}{2} \left\{ f_j'(y) + K f(y) \right\}, \; y \; \in L_j. \tag{4.178}$$

This, along with (4.172), suggest that

$$2 \; h_j(y) = f_j'(y) + K f_j(y). \tag{4.179}$$

Thus, if the conditions in (4.163) are utilized, we observe that the behaviour of $h_j(t)$ at the end points of L_j $(j = 1, 2, 3)$ is as follows:

$$h_1(t) = \begin{cases} O(|t-a|^{-\frac{1}{2}}) & \text{as } t \to a \\ 0 & \text{as } t \to \infty, \end{cases}$$

$$h_2(t) = \begin{cases} O(|t-b|^{-\frac{1}{2}}) & \text{as } t \to b \\ \text{bounded} & \text{as } t \to 0, \end{cases}$$

$$h_3(t) = \begin{cases} O(|t-c|^{-\frac{1}{2}}) & \text{as } t \to c \\ O(|t-d|^{-\frac{1}{2}} & \text{as } t \to d. \end{cases} \tag{4.180}$$

The conditions (4.180) now settle the end conditions to be met by the solutions of the singular integral equations (4.174), which are determined by using the results available in Muskhelishvili (1963). Now we find that

$$h_1(t) = \frac{D_1}{\sqrt{t^2 - a^2}} \quad \text{for } t \in L_1, \tag{4.181}$$

where D_1 is an arbitrary constant,

$$h_2(t) = \frac{C_2\, t}{\pi \sqrt{b^2 - t^2}} \quad \text{for } t \in L_2. \tag{4.182}$$

and

$$\begin{aligned} h_3(t) &= -\frac{1}{\pi \sqrt{(t^2 - c^2)(d^2 - t^2)}} \left[D_3 - C_3 \int_c^d \frac{v\sqrt{(v^2 - c^2)(d^2 - v^2)}\, dv}{v^2 - t^2} \right], \\ &= \frac{C_3}{2} \frac{d_0^2 - t^2}{X(t)}, \quad \text{for } t \in L_3, \end{aligned} \tag{4.183}$$

with

$$\begin{aligned} d_0^2 &= \frac{2}{\pi C_3} \left\{ \frac{\pi C_3}{4} (d^2 + c^2) - D_3 \right\}, \\ X(t) &= \sqrt{(t^2 - c^2)(d^2 - t^2)}, \end{aligned} \tag{4.184}$$

so that $h_3(t)$ contains two unknown constants, viz. d_0^2 and C_3.

Substituting the expressions (4.181)–(4.183) for $h_j(t)$ ($j = 1, 2, 3$) in (4.173), using the results of appropriate standard integrals, we find that, after using (4.171),

$$A_1(k) = D_1 \frac{J_0(ka)}{k^2 + K^2}, \quad A_2(k) = -\frac{b\, C_2}{2} \frac{J_1(kb)}{k^2 + K^2}$$

163

and

$$A_3(k) = \frac{C_3}{\pi} \frac{J(k)}{k^2 + K^2},\tag{4.185}$$

where J_0, J_1 are Bessel functions, and

$$J(k) = \int_c^d \frac{d_0^2 - y^2}{X(y)}\ \sin ky\ dy,\tag{4.186}$$

d_0^2 being determined shortly.

Substituting for A_j for $j = 1, 2, 3$ from (4.185) into (4.166), we finally obtain

$$\begin{aligned} R_1 &= D_1\ K_0(Ka)\ \text{with}\ D_1 = \frac{1}{K_0(Ka) + i\pi I_0(Ka)}, \\ R_2 &= \frac{bC_2}{2}\pi\ I_1(Kb)\ \text{with}\ C_2 b = \frac{2}{\pi\ I_1(Kb) + iK_1(Kb)} \end{aligned}$$

and

$$R_3 = -C_3\gamma_0\ \text{with}\ C_3 = \frac{i}{\alpha_0 - \beta_0 - i\gamma_0},\tag{4.187}$$

where

$$\alpha_0, \beta_0, \gamma_0 = \left(\int_{-c}^c, \int_d^\infty, \int_c^d \right) \frac{(d_0^2 - u^2)\ e^{-Ku}}{|X(u)|}\ du,$$

with

$$d_0^2 = \frac{\displaystyle\int_c^d \frac{u^2 e^{Ku}}{X(u)}\ du}{\displaystyle\int_c^d \frac{e^{Ku}}{X(u)}\ du}.\tag{4.188}$$

Chakrabarti et al. (1997) also considered the general problem, involving an infinite vertical barrier, with a finite number of gaps in it, extending from the surface of deep water to the bottom, by using multiple integral equations of the form (4.14) and (4.15). Recently, Banerjea and Mandal (1998) utilized the same procedure to solve the problem of water wave scattering by a submerged thin vertical wall with a gap.

4.2 Solutions by using the generalized Mehler–Fock inversion theorem

Use of the Mehler–Fock integral transform to solve some classes of dual integral equations involving trigonometric functions as kernel was first made by Babloian (1964) which are discussed in the treatise by Sneddon (1972). In this section, we present some classes of dual integral equations with trigonometric function kernel by applying the generalized Mehler–Fock inversion theorem (cf. Mandal and Mandal 1997, p. 47) and some appropriate properties of associated Legendre functions. The pairs of integral equations considered here were studied by Pathak (1978). These are the generalizations of those considered by Babloian (1964).

First we consider the dual integral equations

$$\int_0^\infty f(\tau) \, \cos x\tau \, d\tau = g(x), \ 0 \le x \le a, \tag{4.189}$$

$$\int_0^\infty f(\tau) \left\{ \Gamma \left(\frac{1}{2} - \mu + i\tau \right) \Gamma \left(\frac{1}{2} - \mu - i\tau \right) \right\}^{-1}$$
$$\times \ \operatorname{cosech} \pi\tau \ \sin x\tau \, d\tau = h(x), \ a < x < \infty. \tag{4.190}$$

To solve this pair, we multiply (4.189) by

$$\sqrt{\frac{2}{\pi}} \left\{ \Gamma \left(\frac{1}{2} - \mu \right) \right\}^{-1} \sinh^\mu x \, (\cosh x - \cosh t)^{-\frac{1}{2} - \mu},$$

integrate it with respect to t from 0 to x and then use the integral representation (A.2.8) to get

$$\int_0^\infty f(\tau) \, P_{-\frac{1}{2} + i\tau}^\mu (\cosh x) \, d\tau = G(x), \ 0 \le x \le a, \tag{4.191}$$

where

$$G(x) = \sqrt{\frac{2}{\pi}} \left\{ \Gamma \left(\frac{1}{2} - \mu \right) \right\}^{-1} \sinh^\mu x \int_0^x g(t) \, (\cosh x - \cosh t)^{-\frac{1}{2} - \mu} \, dt. \tag{4.192}$$

Again, multiplying equation (4.190) by

$$\sqrt{2\pi} \left\{ \Gamma \left(\frac{1}{2} + \mu \right) \right\}^{-1} \sinh^{-\mu} x \, (\cosh t - \cosh x)^{-\frac{1}{2} + \mu},$$

integrating it with respect to t from x to ∞ and then applying the representation (A.2.9), we obtain

$$\int_0^\infty f(\tau) P_{-\frac{1}{2}+i\tau}^\mu (\cosh x) \, d\tau = H(x), \quad a < x < \infty, \tag{4.193}$$

where

$$H(t) = \sqrt{2\pi} \left\{ \Gamma \left(\frac{1}{2} + \mu \right) \right\}^{-1} \sinh^{-\mu} x \int_x^\infty h(t) \, (\cosh t - \cosh x)^{-\frac{1}{2}+\mu} \, dt. \tag{4.194}$$

By the generalized Mehler−Fock inversion theorem, equations (4.191) and (4.193) produce the solution of the pair (4.189) and (4.190) as given by

$$\begin{aligned}
f(\tau) \;=\; & \pi^{-1}\tau \, \sinh \pi\tau \, \Gamma \left(\frac{1}{2} - \mu + i\tau \right) \, \Gamma \left(\frac{1}{2} - \mu - i\tau \right) \\
& \times \left\{ \int_0^a G(x) P_{-\frac{1}{2}+i\tau}^\mu (\cosh x) \, \sinh x \, dx \right. \\
& \left. + \int_a^\infty H(x) P_{-\frac{1}{2}+i\tau}^\mu (\cosh x) \sinh x \, dx \right\}.
\end{aligned} \tag{4.195}$$

Next, we consider the pair

$$\int_0^\infty \tau^{-1} \, f(\tau) \, \sin x\tau \, d\tau = g_1(x), \quad 0 \le x \le a, \tag{4.196}$$

$$\int_0^\infty f(\tau) \left\{ \Gamma \left(\frac{1}{2} - \mu + i\tau \right) \, \Gamma \left(\frac{1}{2} - \mu - i\tau \right) \right\}^{-1}$$

$$\times \quad \operatorname{cosech} \pi\tau \, \sin x\tau \, d\tau = h(x), \quad a < x < \infty. \tag{4.197}$$

The solution of this pair is given by the relation (4.195) where

$$G(x) = \sqrt{\frac{2}{\pi}} \left\{ \Gamma \left(\frac{1}{2} - \mu \right) \right\}^{-1} \sinh^\mu x \int_0^x g_1'(t) \, (\cosh x - \cosh t)^{-\frac{1}{2}-\mu} \, dt, \tag{4.198}$$

instead of the relation (4.192). The function $H(x)$ is defined by the relation (4.194).

Again, we consider another pair

$$\int_0^\infty \tau \, f(\tau) \, \sin x\tau \, d\tau = g_2(x), \quad 0 \le x \le a, \tag{4.199}$$

$$\int_0^\infty f(\tau) \left\{ \Gamma \left(\frac{1}{2} - \mu + i\tau \right) \, \Gamma \left(\frac{1}{2} - \mu - i\tau \right) \right\}^{-1}$$

$$\times \quad \operatorname{cosech} \pi\tau \, \sin x\tau \, d\tau = h(x), \quad a < x < \infty. \tag{4.200}$$

166

To obtain the solution of this pair, we integrate (4.199) with respect to x from 0 to x. Then equation (4.199) assumes the form (4.189) with $g(x)$ now defined by

$$g(x) = A - \int_0^x g_2(t)\, dt, \tag{4.201}$$

where the constant A is defined by

$$A = \int_0^\infty f(\tau)\, d\tau. \tag{4.202}$$

The solution of this pair is then obtained in terms of A by the relations (4.195), (4.192) and (4.194) with $g(x)$ defined by the relation (4.201).

The solution of the dual integral equations

$$\int_0^\infty f(\tau)\, \cos x\tau\, d\tau = g_1(x),\ 0 \le x \le a, \tag{4.203}$$

$$\int_0^\infty \tau^{-1}\, f(\tau) \left\{ \Gamma\left(\frac{1}{2} - \mu + i\tau\right)\, \Gamma\left(\frac{1}{2} - \mu - i\tau\right) \right\}^{-1}$$
$$\times\ \ \text{cosech}\,\pi\tau\,\ \cos x\tau\, d\tau = h_1(x),\ a < x < \infty, \tag{4.204}$$

is given by the relations (4.195), (4.192) and (4.194) with $h(x)$ now defined by

$$h(x) = -h_1'(x).$$

Finally, the solution of the pair of integral equations

$$\int_0^\infty f(\tau)\, \cos x\tau\, d\tau = g(x),\ 0 \le x \le a, \tag{4.205}$$

$$\int_0^\infty f(\tau) \left\{ \Gamma\left(\frac{1}{2} - \mu + i\tau\right)\, \Gamma\left(\frac{1}{2} - \mu - i\tau\right) \right\}^{-1}$$
$$\times\ \ \text{cosech}\,\pi\tau\,\ \cos x\tau\, d\tau = h_2(x),\ a < x < \infty. \tag{4.206}$$

is obtained from the relations (4.195), (4.192) and (4.194) with $h(x)$ now defined by

$$h(x) = A - \int_x^\infty h_2(x)\, dt.$$

The dual integral equations together with their solutions studied by Babloian (1964) can be deduced as special cases of the results presented here by putting $\mu = 0$ and using the facts that

$$\Gamma\left(\frac{1}{2} + i\tau\right) \Gamma\left(\frac{1}{2} - i\tau\right) = \pi\, \text{sech}\,\pi\tau \ \text{ and } \ P^0_{-\frac{1}{2}+i\tau}(x) = P_{-\frac{1}{2}+i\tau}(x).$$

4.3 Solutions by using the generalized associated Mehler−Fock inversion theorem

In this section, we consider some classes of dual integral equations involving trigonometric function kernels whose solutions are obtained by using the generalized associated Mehler−Fock inversion theorem (cf. Mandal and Mandal 1997, p. 58). These are generalizations of earlier ones considered in Section 4.2. These pairs were investigated by N. Mandal (1995).

First we consider the dual integral equations

$$\int_0^\infty f(\tau)\ \cos x\tau\ d\tau\ =\ g(x),\ 0 \le x \le a, \tag{4.207}$$

$$\int_0^\infty \omega(\tau)\ f(\tau)\ \sin x\tau\ d\tau\ =\ h(x),\ a < x < \infty, \tag{4.208}$$

where

$$\omega(\tau) = \ \operatorname{cosech} 2\pi\tau \left\{ \Gamma\left(\frac{1-\mu+\nu}{2}+i\tau\right)\ \Gamma\left(\frac{1-\mu+\nu}{2}-i\tau\right) \Gamma\left(\frac{1-\mu-\nu}{2}+i\tau\right)\right.$$
$$\left. \times\Gamma\left(\frac{1-\mu-\nu}{2}-i\tau\right)\right\}^{-1}$$

is a known weight function.

To solve this pair, we multiply (4.207) by

$$\pi^{-\frac{1}{2}}2^{(\mu-\nu-\frac{1}{2})/2}\left\{\Gamma\left(\frac{1}{2}-\mu\right)\right\}^{-1}\sinh^\mu x(\cosh x - \cosh t)^{-\frac{1}{2}-\mu}$$

$$\times F\left(\frac{\nu-\mu}{2},-\frac{\nu+\mu}{2};\ \frac{1}{2}-\mu;\ \frac{\cosh x - \cosh t}{1+\cosh x}\right), \tag{4.209}$$

and integrate it with respect to t from 0 to x, then interchange the order of integration and apply the representation (A.3.2) to obtain

$$\int_0^\infty f(\tau)\ P_{-\frac{1}{2}+i\tau}^{\mu,\nu}(\cosh x)\ d\tau = G_1(x),\ 0 \le x \le a, \tag{4.210}$$

where

$$G_1(x) = \pi^{-\frac{1}{2}} 2^{(\mu-\nu-1)/2} \left\{ \Gamma\left(\frac{1}{2} - \mu\right) \right\}^{-1} \sinh^{\mu} x \int_0^x (\cosh x - \cosh t)^{-\frac{1}{2}-\mu}$$

$$\times F\left(\frac{\nu-\mu}{2}, -\frac{\nu+\mu}{2}; \frac{1}{2} - \mu; \frac{\cosh x - \cosh t}{1 + \cosh x}\right) g(t) \, dt. \tag{4.211}$$

Also, if we multiply (4.208) by

$$(2\pi)^{\frac{3}{2}} 2^{(\nu-\mu)/2} \left\{ \Gamma\left(\frac{1}{2} + \mu\right) \right\}^{-1} \sinh^{-\mu} x (\cosh t - \cosh x)^{-\frac{1}{2}+\mu}$$

$$\times F\left(\frac{\mu-\nu}{2}, \frac{\mu+\nu}{2}; \frac{1}{2} + \mu; \frac{\cosh x - \cosh t}{1 + \cosh x}\right), \tag{4.212}$$

integrate it with respect to t from x to ∞, then interchange the order of integration and then use the representation (A.3.14), we get

$$\int_0^{\infty} f(\tau) \, P^{\mu,\nu}_{-\frac{1}{2}+i\tau}(\cosh x) \, d\tau = H_1(x), \ a < x < \infty, \tag{4.213}$$

where

$$H_1(x) = (2\pi)^{\frac{3}{2}} 2^{(\nu-\mu)/2} \left\{ \Gamma\left(\frac{1}{2} + \mu\right) \right\}^{-1} \sinh^{-\mu} x \int_x^{\infty} (\cosh t - \cosh x)^{-\frac{1}{2}+\mu}$$

$$\times F\left(\frac{\mu-\nu}{2}, \frac{\mu+\nu}{2}; \frac{1}{2} + \mu; \frac{\cosh x - \cosh t}{1 + \cosh x}\right) h(t) \, dt. \tag{4.214}$$

Thus by the generalized associated Mehler−Fock inversion theorem, we obtain the solution of the pair (4.207) and (4.208), from (4.210) and (4.213), given by

$$f(\tau) = \frac{2^{\mu-\nu-1}}{\pi^2 \omega(\tau)} \left[\int_0^a G_1(x) \, P^{\mu,\nu}_{-\frac{1}{2}+i\tau}(\cosh x) \, \sinh x \, dx \right.$$

$$\left. + \int_a^{\infty} H_1(x) \, P^{\mu,\nu}_{-\frac{1}{2}+i\tau}(\cosh x) \sinh x \, dx \right]. \tag{4.215}$$

Next, the solution of the pair

$$\int_0^{\infty} \tau^{-1} \, f(\tau) \, \sin x\tau \, d\tau = m(x), \ 0 \le x \le a, \tag{4.216}$$

$$\int_0^{\infty} \omega(\tau) \, f(\tau) \, \sin x\tau \, d\tau = h(x), \ a < x < \infty, \tag{4.217}$$

is given by the relation (4.215) where we define

169

$$G_1(x) = \pi^{-\frac{1}{2}} 2^{(\mu-\nu-1)/2} \left\{ \Gamma\left(\frac{1}{2} - \mu\right) \right\}^{-1} \sinh^\mu x \int_0^x (\cosh x - \cosh t)^{-\frac{1}{2}-\mu}$$

$$\times F\left(\frac{\nu - \mu}{2}, -\frac{\nu + \mu}{2}; \frac{1}{2} - \mu; \frac{\cosh x - \cosh t}{1 + \cosh x}\right) m'(x) \, dx, \tag{4.218}$$

instead of (4.211). The function $H_1(x)$ is the same as that defined by (4.214).

The solution of the pair of integral equations

$$\int_0^\infty \tau \, f(\tau) \, \sin x\tau \, d\tau = n(x), \; 0 \le x \le a, \tag{4.219}$$

$$\int_0^\infty \omega(\tau) \, f(\tau) \, \sin x\tau \, d\tau = h(x), \; a < x < \infty, \tag{4.220}$$

is obtained by integrating the equation (4.219) with respect to x from 0 to x. It then assumes the form (4.207) with $g(x)$ now defined by

$$g(x) = C - \int_0^x n(u) \, du, \tag{4.221}$$

where C is a constant defined by

$$C = \int_0^\infty f(\tau) \, d\tau. \tag{4.222}$$

The solution is then given in terms of C by (4.215), (4.211) and (4.214) with $g(x)$ defined by (4.221). C is ultimately obtained by using the relation (4.222).

The solution of the pair

$$\int_0^\infty f(\tau) \, \cos x\tau \, d\tau = g(x), \; 0 \le x \le a, \tag{4.223}$$

$$\int_0^\infty \tau^{-1} \omega(\tau) \, f(\tau) \, \cos x\tau \, d\tau = p(x), \; a < x < \infty, \tag{4.224}$$

is given by the relations (4.215), (4.211) and (4.214) with $h(x)$ now defined by

$$h(x) = -p'(x). \tag{4.225}$$

Finally, the solution of the dual integral equations

$$\int_0^\infty f(\tau) \, \cos x\tau \, d\tau = g(x), \; 0 \le x \le a, \tag{4.226}$$

$$\int_0^\infty \tau \, \omega(\tau) \, f(\tau) \, \cos x\tau \, d\tau = q(x), \; a < x < \infty, \tag{4.227}$$

is given by the relations (4.215), (4.211) and (4.214) with $h(x)$ now defined by

$$h(x) = C_1 - \int_a^\infty q(u) \ du. \qquad (4.228)$$

The unknown constant C_1 is determined in the same way as in (4.222).

It may be mentioned here that all the dual integral equations and their solutions presented in Section 4.2 can be deduced as special cases of the results obtained in this section simply on setting $\mu = \nu$ and using the fact that

$$P^{\mu,\mu}_{-\frac{1}{2}+i\tau}(\cosh x) = P^{\mu}_{-\frac{1}{2}+i\tau}(\cosh x).$$

Chapter 5

Dual integral equations involving inverse Mellin transforms

In this chapter, we consider the solutions to some dual integral equations involving inverse Mellin transforms. These dual integral equations arise in the solutions of certain mixed boundary value problems of potential theory for wedge-shaped regions. Srivastav (1965) considered the solution of Laplace's equation

$$\nabla^2 \phi \equiv \frac{\partial^2 \phi}{\partial \rho^2} + \frac{1}{\rho} \frac{\partial \phi}{\partial \rho} + \frac{1}{\rho^2} \frac{\partial^2 \phi}{\partial \theta^2} + \frac{\partial^2 \phi}{\partial z^2} = 0, \tag{5.1}$$

for wedge-shaped regions subject to mixed boundary conditions. If ϕ is assumed to be independent of z, then equation (5.1) is reduced to the form

$$\frac{\partial^2 \phi}{\partial \rho^2} + \frac{1}{\rho} \frac{\partial \phi}{\partial \rho} + \frac{1}{\rho^2} \frac{\partial^2 \phi}{\partial \theta^2} = 0. \tag{5.2}$$

The wedge is supposed to occupy the region $0 < \rho < \infty$, $0 \leq \theta \leq \gamma$. On the face $\theta = \gamma$,

either (a) $\phi(\rho, \theta) = \phi_0(\rho)$,

or (b) $\dfrac{\partial \phi}{\partial \rho} = \phi_1(\rho)$,

where $\phi_0(\rho)$ or $\phi_1(\rho)$ are known functions. Corresponding to each of these two conditions, there may be two kinds of boundary conditions on the face $\theta = 0$:

(i) $\phi(\rho, 0) = f_1(\rho), \ 0 < \rho < 1,$

$\dfrac{\partial \phi}{\partial \rho}\Big|_{\theta=0} = f_2(\rho), \ \rho > 1;$

(ii) $\dfrac{\partial \phi}{\partial \theta}\Big|_{\theta=0} = f_1(\rho), \ 0 < \rho < 1,$

$\phi(\rho, 0) = f_2(\rho), \ \rho > 1,$

where $f_1(\rho)$ and $f_2(\rho)$ are known functions.

Applying the Mellin integral transform, a solution of (5.2) in terms of an auxiliary function ψ satisfying boundary condition (a), is obtained as

$$\phi(\rho, \theta) = \frac{1}{2\pi i} \int_{c-i\infty}^{c+i\infty} \psi(s) \ \sec \alpha s \ \sin \left[(\gamma - \theta) s \right] \rho^{-s} \ ds$$

$$+ \frac{1}{2\pi i} \int_{c_1-i\infty}^{c_1+i\infty} \overline{\phi}_0(s) \ \sec \gamma s \ \cos \left[(\gamma - \theta) s \right] \rho^{-s} \ ds, \qquad (5.3)$$

where $\overline{\phi}_0(s)$ is the Mellin transform of $\phi_0(\rho)$, and a solution of (5.2) satisfying boundary condition (b) is given by

$$\phi(\rho, \theta) = \frac{1}{2\pi i} \int_{c-i\infty}^{c+i\infty} \psi(s) \ \sec \gamma s \ \cos \left[(\gamma - \theta) s \right] \rho^{-s} \ ds$$

$$- \frac{1}{2\pi i} \int_{c_2-i\infty}^{c_2+i\infty} s^{-1} \ \overline{\phi}_1(s) \ \sec \gamma s \ \sin \left[(\gamma - \theta) s \right] \rho^{-s} \ ds, \qquad (5.4)$$

where $\overline{\phi}_1(s)$ is the Mellin transform of $\phi_1(\rho)$. The constants c, c_1, c_2 are to be chosen suitably so that the functions $\rho^{c-1}\phi(\rho, \theta)$, $\rho^{c_1-1} \phi_0(\rho)$, $\rho^{c_2-1} \phi_1(\rho)$ and $\psi(c + i\sigma)$ are integrable and have bounded variation.

The boundary conditions on $\theta = 0$ produce the following four types of dual integral equations to determine the auxiliary function $\psi(s)$ in each case:

(i) $\dfrac{1}{2\pi i} \displaystyle\int_{c-i\infty}^{c+i\infty} \psi(s) \ \tan \gamma s \ \rho^{-s} \ ds = f_1(\rho), \ 0 < \rho < 1,$

$\dfrac{1}{2\pi i} \displaystyle\int_{c-i\infty}^{c+i\infty} s \ \psi(s) \ \rho^{-s} \ ds = -f_2(\rho), \ \rho > 1;$

(ii)
$$\frac{1}{2\pi i} \int_{c-i\infty}^{c+i\infty} s\,\psi(s)\,\rho^{-s}\,ds \;=\; -f_1(\rho),\; 0 < \rho < 1,$$
$$\frac{1}{2\pi i} \int_{c-\infty}^{c+i\infty} \psi(s)\,\tan\gamma s\,\rho^{-s}\,ds \;=\; f_2(\rho),\; \rho > 1;$$

(iii)
$$\frac{1}{2\pi i} \int_{c-i\infty}^{c+i\infty} \psi(s)\,\rho^{-s}\,ds \;=\; f_1(\rho),\; 0 < \rho < 1,$$
$$\frac{1}{2\pi i} \int_{c-i\infty}^{c+i\infty} s\,\psi(s)\,\tan\gamma s\,\rho^{-s}\,ds \;=\; f_2(\rho),\; \rho > 1;$$

(iv)
$$\frac{1}{2\pi i} \int_{c-i\infty}^{c+i\infty} s\,\psi(s)\,\tan\gamma s\,\rho^{-s}\,ds \;=\; f_1(\rho),\; 0 < \rho < 1,$$
$$\frac{1}{2\pi i} \int_{c-i\infty}^{c+i\infty} \psi(s)\,\rho^{-s}\,ds \;=\; f_2(\rho),\; \rho > 1.$$

The use of the inversion theorem for the Mellin transform (cf. Titchmarsh 1937, p. 46) and the following two elementary integrals

$$\int_0^t x^{-s}\,(t^2 - x^2)^{-\frac{1}{2}}dx \;=\; \frac{\Gamma(\frac{1}{2} - \frac{1}{2}s)\Gamma(\frac{1}{2})}{2\Gamma(1 - \frac{s}{2})}\,t^{-s},\; Re(s) < \frac{1}{2},$$

$$\int_t^\infty x^{-s}\,(x^2 - t^2)^{-\frac{1}{2}}dx \;=\; \frac{\Gamma(\frac{s}{2})\,\Gamma(\frac{1}{2})}{2\Gamma(\frac{1}{2} + \frac{s}{2})}\,t^{-s},\; Re(s) > 0,$$

ultimately reduce each pair of integral equations to a single Fredholm integral equation of the second kind. The details of the calculations are available in the paper of Srivastav (1965).

However, the explicit solution to these dual integral equations has been obtained by Srivastav and Parihar (1968) (also see Virchenko 1989, p. 100). The following results are most useful to obtain explicit solutions to these pairs.

The solution of the Abel type integral equation of the first kind

$$\int_0^x g(t)\,(x^\alpha - t^\alpha)^{-\frac{1}{2}}\,dt = f(x),\; 0 < x < 1,\; \alpha > 0, \tag{5.5}$$

is given by

$$g(t) = \frac{\alpha}{\pi}\,\frac{d}{dt} \int_0^t x^{\alpha-1}\,(t^\alpha - x^\alpha)^{-\frac{1}{2}}\,f(x)\,dx, \tag{5.6}$$

and the solution of the equation

$$\int_x^\infty g(t)\,(t^\alpha - x^\alpha)^{-\frac{1}{2}}\,dt = f(x),\; 1 < x < \infty,\; \alpha > 0, \tag{5.7}$$

is given by

$$g(t) = -\frac{\alpha}{\pi} \frac{d}{dt} \int_t^\infty x^{\alpha-1} \left(x^\alpha - t^\alpha\right)^{-\frac{1}{2}} f(x) \, dx. \tag{5.8}$$

It follows from Mellin's inversion theorem and integral relations (cf. Erdélyi et al. 1954a, p. 311) that, for $c > 0$

$$\frac{1}{2\pi i} \int_{c-i\infty}^{c+i\infty} \alpha^{-1} B\left(\frac{1}{2}, \frac{s}{\alpha}\right) x^{-s} \, ds = \begin{cases} (1 - x^\alpha)^{-\frac{1}{2}}, & 0 < x < 1, \\ 0, & x > 1; \end{cases} \tag{5.9}$$

and that, for $c < \frac{\alpha}{2}$

$$\frac{1}{2\pi i} \int_{c-i\infty}^{c+i\infty} \alpha^{-1} B\left(\frac{1}{2}, \frac{1}{2} - \frac{s}{\alpha}\right) x^{-s} \, ds = \begin{cases} 0, & 0 < x < 1, \\ (x^\alpha - 1)^{-\frac{1}{2}}, & x > 1, \end{cases} \tag{5.10}$$

where $B(x, y)$ denotes the Beta function.

First we consider the dual integral equations

$$\frac{1}{2\pi i} \int_{c-i\infty}^{c+i\infty} \psi(s) \tan \gamma s \, \rho^{-s} \, ds = f_1(\rho), \; 0 < \rho < 1, \tag{5.11}$$

$$\frac{1}{2\pi i} \int_{c-i\infty}^{c+i\infty} s \, \psi(s) \, \rho^{-s} \, ds = f_2(\rho), \; \rho > 1. \tag{5.12}$$

To solve this pair, first we assume that $f_2(\rho) = 0$ and let

$$\frac{1}{2\pi i} \int_{c-i\infty}^{c+i\infty} s \, \psi(s) \, \rho^{-s} \, ds = -\rho \, \frac{\partial}{\partial \rho} \int_\rho^1 \psi_1(t) \, (t^\alpha - \rho^\alpha)^{-\frac{1}{2}} \, dt, \; 0 < \rho < 1, \tag{5.13}$$

where $\psi_1(t)$ is an unknown function to be determined and $\alpha = \frac{\pi}{\gamma}$. By Mellin's integral transform theorem, (5.13) produces

$$\psi(s) = \int_0^1 \psi_1(t) \left\{ \int_0^t \rho^{s-1} \, (t^\alpha - \rho^\alpha)^{-\frac{1}{2}} \, d\rho \right\} dt. \tag{5.14}$$

Then using the relation (5.9) and properties of the gamma functions, (5.14) reduces to

$$\psi(s) \tan \gamma s = \frac{1}{\alpha} B\left(\frac{1}{2}, \frac{1}{2} - \frac{s}{\alpha}\right) \int_0^1 t^{s - \frac{\alpha}{2}} \psi_1(t) \, dt. \tag{5.15}$$

Substituting this value of $\psi(s) \tan \gamma s$ from (5.15) into the relation (5.11), we get an Abel type integral equation

$$\int_0^\rho \psi_1(t) \, (\rho^\alpha - t^\alpha)^{-\frac{1}{2}} \, dt = f_1(\rho), \; 0 < \rho < 1, \tag{5.16}$$

whose solution is (cf. (5.6))

$$\psi_1(t) = \frac{\alpha}{\pi} \frac{d}{dt} \int_0^t \rho^{\alpha-1} f_1(\rho) \ (t^\alpha - \rho^\alpha)^{-\frac{1}{2}} \ d\rho. \tag{5.17}$$

Thus, the solution of the pair (5.11), (5.12) with $f_2(\rho) = 0$ is given by (5.14) where $\psi_1(t)$ is obtained by (5.17).

Next, we suppose that $f_1(\rho) = 0$ and let

$$\frac{1}{2\pi i} \int_{c-i\infty}^{c+i\infty} \psi(s) \ \tan \gamma s \ \rho^{-s} \ ds = \int_1^\rho \psi_2(t) \ (\rho^\alpha - t^\alpha)^{-\frac{1}{2}} \ dt, \ \rho > 1. \tag{5.18}$$

Using Mellin's integral transform theorem and the relation (5.10), from (5.18) we obtain

$$\psi(s) = \frac{1}{\alpha} \ B\left(\frac{1}{2}, \frac{s}{\alpha}\right) \int_1^\infty t^{s-\frac{\alpha}{2}} \ \psi_2(t) \ dt. \tag{5.19}$$

Equation (5.12) can be rewritten as

$$\rho \frac{\partial}{\partial \rho} \frac{1}{2\pi i} \int_{c-i\infty}^{c+i\infty} \psi(s) \ \rho^{-s} \ ds = -f_2(\rho), \ \rho > 1. \tag{5.20}$$

Then, from (5.19) and (5.20), we get an Abel type integral equation

$$\frac{\partial}{\partial \rho} \int_\rho^\infty \psi_2(t) \ (t^\alpha - \rho^\alpha)^{-\frac{1}{2}} dt = -\frac{f_2(\rho)}{\rho}, \ \rho > 1, \tag{5.21}$$

whose solution is (cf. (5.8))

$$\psi_2(t) = \frac{\alpha}{\pi} \ t^{\alpha-1} \int_t^\infty \rho^{-1} f_2(\rho) \ (\rho^\alpha - t^\alpha)^{-\frac{1}{2}} \ d\rho. \tag{5.22}$$

Hence, the solution of the pair (5.11) and (5.12) is given by

$$\psi(s) = \frac{1}{\alpha} \ B\left(\frac{1}{2}, \frac{s}{\alpha}\right) \int_0^\infty \Psi(t) \ t^{s-\frac{\alpha}{2}} \ dt, \tag{5.23}$$

where

$$\Psi(t) = \begin{cases} \psi_1(t), & 0 < t < 1, \\ \psi_2(t), & t > 1. \end{cases}$$

In a similar manner, it can be shown that the solution of the dual integral equations

$$\frac{1}{2\pi i} \int_{c-i\infty}^{c+i\infty} s \ \psi(s) \ \rho^{-s} \ ds = f_1(\rho), \ 0 < \rho < 1, \tag{5.24}$$

$$\frac{1}{2\pi i} \int_{c-i\infty}^{c+i\infty} \psi(s) \ \tan \alpha s \ \rho^{-s} \ ds = f_2(\rho), \ \rho > 1, \tag{5.25}$$

176

is given by

$$\psi(s) = \frac{1}{\alpha} B\left(\frac{1}{2}, -\frac{s}{\alpha}\right) \int_0^\infty \Psi(t) \, t^s \, dt, \qquad (5.26)$$

where

$$\Psi(t) = \begin{cases} \dfrac{\alpha}{\pi t} \displaystyle\int_0^t \rho^{\frac{\alpha}{2}-1} \, (t^\alpha - \rho^\alpha)^{-\frac{1}{2}} \, f_1(\rho) \, d\rho, & 0 < t < 1 \\[4mm] \dfrac{\alpha}{\pi} \dfrac{d}{dt} \displaystyle\int_t^\infty \rho^{\frac{\alpha}{2}-1} \, (\rho^\alpha - t^\alpha)^{-\frac{1}{2}} \, f_2(\rho) \, d\rho, & t > 1. \end{cases}$$

Similarly, the solution of the pair

$$\frac{1}{2\pi i} \int_{c-i\infty}^{c+i\infty} \psi(s) \, \rho^{-s} \, ds \;=\; f_1(\rho), \; 0 < \rho < 1, \qquad (5.27)$$

$$\frac{1}{2\pi i} \int_{c-i\infty}^{c+i\infty} s \, \psi(s) \, \tan \alpha s \, \rho^{-s} \, ds \;=\; f_2(\rho), \; \rho > 1, \qquad (5.28)$$

is given by

$$\psi(s) = \frac{1}{\alpha} B\left(\frac{1}{2}, -\frac{s}{\alpha}\right) \int_0^\infty \Psi(t) \, t^s \, dt, \qquad (5.29)$$

where

$$\Psi(t) = \begin{cases} \dfrac{\alpha}{\pi} \dfrac{d}{dt} \displaystyle\int_0^t \rho^{\frac{\alpha}{2}-1} \, (t^\alpha - \rho^\alpha)^{-\frac{1}{2}} \, f_1(\rho) \, d\rho, & 0 < t < 1 \\[6mm] -\dfrac{\alpha}{\pi} t^{\frac{\alpha}{2}-1} \displaystyle\int_t^\infty \rho^{-1} \, f_2(\rho) \, (\rho^\alpha - t^\alpha)^{-\frac{1}{2}} \, d\rho, & t > 1. \end{cases}$$

Finally, the solution of the dual integral equations

$$\frac{1}{2\pi i} \int_{c-i\infty}^{c+i\infty} s \, \psi(s) \, \tan \alpha s \, \rho^{-s} \, ds \;=\; f_1(\rho), \; 0 < \rho < 1, \qquad (5.30)$$

$$\frac{1}{2\pi i} \int_{c-i\infty}^{c+i\infty} \psi(s) \, \rho^{-s} \, ds \;=\; f_2(\rho), \; \rho > 1, \qquad (5.31)$$

may be expressed by the relation

$$\psi(s) = \frac{1}{\alpha} B\left(\frac{1}{2}, \frac{s}{\alpha}\right) \int_0^\infty t^{s-\frac{\alpha}{2}} \, \Psi(t) dt, \qquad (5.32)$$

where

$$\Psi(t) = \begin{cases} -\dfrac{\alpha}{\pi} t^{\alpha-1} \displaystyle\int_0^t \rho^{-1} \, (t^\alpha - \rho^\alpha)^{-\frac{1}{2}} \, f_1(\rho) \, d\rho, & 0 < t < 1 \\[6mm] -\dfrac{\alpha}{\pi} \dfrac{d}{dt} \displaystyle\int_t^\infty \rho^{\alpha-1} \, (\rho^\alpha - t^\alpha)^{-\frac{1}{2}} \, f_2(\rho) \, d\rho, & t > 1. \end{cases}$$

The solutions of the above dual integral equations can also be obtained by using some fractional integrals. How a systematic use of fractional integrals leads in a

177

simple manner to the solutions of the dual integral equations discussed above as well as to the solutions of some further integral equations is presented below (cf. Erdélyi 1968).

First we denote the Mellin transform of a function $f(x)$ by

$$F(s) = M[f(x)] = \int_0^\infty f(x) \, x^{s-1} \, dx. \tag{5.33}$$

We consider Kober's fractional integration operators which are slightly extended to integration with respect to x^n (n is a positive number) rather than x as

$$I_{x^n}^{\eta,\alpha} f(x) = \frac{n}{\Gamma(\alpha)} \, x^{-n\alpha-n\eta} \int_0^x (x^n - t^n)^{\alpha-1} \, f(t) \, t^{n\eta+n-1} \, dt, \tag{5.34}$$

$$K_{x^n}^{\eta,\alpha} f(x) = \frac{n}{\Gamma(\alpha)} \, x^{n\eta} \int_x^\infty (t^n - x^n)^{\alpha-1} \, f(t) \, t^{-n\alpha-n\eta+n-1} \, dt. \tag{5.35}$$

Then it is obtained that

$$M[I_{x^n}^{\eta,\alpha} f(x)] = \frac{\Gamma(1+\eta-\frac{s}{n})}{\Gamma(1+\eta+\alpha-\frac{s}{n})} \, M[f(x)], \tag{5.36}$$

$$M[K_{x^n}^{\eta,\alpha} f(x)] = \frac{\Gamma(\eta+\frac{s}{n})}{\Gamma(\eta+\alpha+\frac{s}{n})} \, M[f(x)]. \tag{5.37}$$

Let us assume two functions $f_1(x)$ and $f_2(x)$ which are only partially known such that $f_1(x)$ is given for $0 < x < 1$ and $f_2(x)$ for $1 < x < \infty$. We also suppose that the Mellin transforms of the functions $f_1(x)$ and $f_2(x)$ satisfy the relation

$$\frac{F_1(s)}{F_2(s)} = \frac{\Gamma(1+\eta+\alpha-\frac{s}{m})\Gamma(\xi+\frac{s}{n})}{\Gamma(1+\eta-\frac{s}{m}) \, \Gamma(\xi+\beta+\frac{s}{n})}, \tag{5.38}$$

where $\alpha, \beta, \xi, \eta, m$ and n are known. Then we wish to find the functions $f_1(x), f_2(x),$ $F_1(s)$ and $F_2(s)$. To obtain these functions, we write

$$F(s) = \frac{\Gamma(1+\eta-\frac{s}{m})}{\Gamma(1+\eta+\alpha-\frac{s}{m})} \, F_1(s) = \frac{\Gamma(\xi+\frac{s}{n})}{\Gamma(\xi+\beta+\frac{s}{n})} \, F_2(s) = M[f(x)]. \tag{5.39}$$

Then, using (5.36) and (5.37), we have

$$f(x) = I_{x^m}^{\eta,\alpha} \, f_1(x) = K_{x^n}^{\alpha,\beta} f_2(x). \tag{5.40}$$

178

From the first relation of (5.40), $f(x)$ can be found for $0 < x < 1$ and from the other relation, $f(x)$ can be obtained for $1 < x < \infty$. Therefore, the function $f(x)$ is known and so the functions $F(s), F_1(s), F_2(s)$ are also known, while

$$f_1(x) = I_{x^m}^{\eta+\alpha,-\alpha} f(x) \text{ and } f_2(x) = K_{x^n}^{\xi+\beta,-\beta} f(x). \tag{5.41}$$

The dual integral equations discussed above correspond to $F_1(s)/F_2(s)$ being, respectively,

$$\left(\frac{n}{s} \tan \frac{\pi s}{n}\right)^{\pm 1}, \left(\frac{s}{n} \tan \frac{\pi s}{n}\right)^{\pm 1}.$$

We see here that

$$\frac{n}{s} \tan \frac{\pi s}{n} = \frac{n}{s} \frac{\sin(\frac{\pi s}{n})}{\cos(\frac{\pi s}{n})} = \frac{n}{s} \frac{\Gamma(\frac{1}{2} + \frac{s}{n})\Gamma(\frac{1}{2} - \frac{s}{n})}{\Gamma(\frac{s}{n})\Gamma(1 - \frac{s}{n})} = \frac{\Gamma(\frac{1}{2} + \frac{s}{n})\Gamma(\frac{1}{2} - \frac{s}{n})}{\Gamma(1 + \frac{s}{n})\Gamma(1 - \frac{s}{n})},$$

and there are corresponding relations in the three other cases. Thus, the solutions of the dual integral equations presented above now follow easily. There are many other cases leading to an expression of (5.38) in terms of elementary functions.

The above technique can also be used if

$$\frac{F_1(s)}{F_2(s)} = \prod_{i=1}^{p} \frac{\Gamma(1 + \eta_i + \alpha_i - \frac{s}{m_i})}{\Gamma(1 + \eta_i - \frac{s}{m_i})} \prod_{j=1}^{q} \frac{\Gamma(\xi_j + \frac{s}{n_j})}{\Gamma(\xi_j + \beta_j + \frac{s}{n_j})},$$

and the result can be expressed either in terms of products of operators of fractional integration, or in terms of operators whose kernel is a generalized hypergometric function.

More generally,

$$\frac{F_1(s)}{F_2(s)} = \frac{G_2(s)}{G_1(s)}$$

can be treated in this manner, if $G_1(s)$ and $G_2(s)$ are given functions which are analytic in a right-half plane and left-half plane respectively.

As applications of the dual integral equations involving inverse Mellin transforms to some physical problems, Srivastav and Narain (1965) considered certain two-dimensional problems of stress distribution in wedge-shaped elastic solids under discontinuous load. They studied the problem of the distribution of stress in an infinite wedge of a homogeneous elastic isotropic solid under the usual assumptions pertaining to plane strain in the classical (infinitesimal) theory of elasticity. The wedge occupies the region $0 \leq \rho < \infty$, $-\alpha \leq \theta \leq \alpha$ (in plane polar coordinates) where the pole is taken on the apex of the wedge with the line bisecting the wedge angle as the initial line. The following two types of stress fields are considered here.

(i) The stress field which is set up by the application of known pressure to inner surfaces of a crack situated on the bisector of the wedge angle.

(ii) The stress field generated by the indentation of the plane faces of the wedge by the rigid punch.

The corresponding boundary value problems can be reduced to the problem of solving dual integral equations involving inverse Mellin transforms. The details of the calculations are available in the paper of Srivastav and Narain (1965).

Now we consider an axisymmetric problem of the torsion of an elastic space, weakened by a conical crack. The solution of this problem is obtained by the application of dual integral equations related to the Mellin integral transform (cf. Zlatina 1972).

Let us assume r, θ, ϕ to be the system of spherical coordinates whose origin coincides with the vertex of the crack and the z-axis is the axis of symmetry. For this choice of coordinate system, the problem under consideration reduces to the determination of the only non-zero component $u_\phi = u(r, \theta)$ of the displacement vector, satisfying the equation

$$\nabla^2 u - \frac{u}{r^2 \sin^2 \theta} = 0, \tag{5.42}$$

in the region $0 \leq r < \infty$, $0 \leq \theta \leq \pi$, $0 \leq \phi < 2\pi$, where

$$\nabla^2 = \frac{1}{r^2} \frac{\partial}{\partial r} \left(r^2 \frac{\partial}{\partial r} \right) + \frac{1}{r^2 \sin \theta} \frac{\partial}{\partial \theta} \left(\sin \theta \frac{\partial}{\partial \theta} \right).$$

The component of the stress tensor can be expressed as

$$\tau_{\theta\phi} = \mu \left[\frac{\partial}{\partial \theta} \left(\frac{u}{\sin \theta} \right) \right] \frac{\sin \theta}{r}, \quad \tau_{r\phi} = \mu \left[\frac{\partial u}{\partial r} - \frac{u}{r} \right], \tag{5.43}$$

where μ is the shear modulus. On the boundaries of the crack, the tangential displacements or the shear stresses are prescribed as given by

$$u|_{\theta=\alpha} = f_1(r), \ 0 \leq r < a, \tag{5.44}$$

or

$$\sin\theta \frac{\partial}{\partial\theta}\left(\frac{u}{\sin\theta}\right)|_{\theta=\alpha} = \frac{1}{\mu} f_2(r), \ 0 \leq r < a, \tag{5.45}$$

where $f_1(r)$ and $f_2(r)$ are known continuous functions. The conditions at infinity and near the vertex of the cone are given by

$$\left.\begin{array}{rcl} u|_{r\to\infty} & = & O(r^{-1}), \ \tau_{r\phi} = O(r^{-2}), \ \tau_{\theta\phi} = O(r^{-2}), \\ u|_{\theta=0,r\to0} & = & O(1), \ \tau_{r\phi} = O(r^{-\delta}), \ 0 < \delta < 1, \\ \tau_{\theta\phi} & = & O(r^{-\epsilon}), \ 0 < \epsilon < 1. \end{array}\right\} \tag{5.46}$$

The solution of the above boundary value problem can be represented in the form

$$u = \begin{cases} \sqrt{\dfrac{a}{r}} \displaystyle\int_{-\infty}^{\infty} f(\tau) \dfrac{P^1_{-\frac{1}{2}+i\tau}(\cos\theta)}{P^1_{-\frac{1}{2}+i\tau}(\cos\alpha)} e^{-i\tau \ln(r/a)} \, d\tau, & 0 \leq \theta < \alpha, \\[4mm] \sqrt{\dfrac{a}{r}} \displaystyle\int_{-\infty}^{\infty} f(\tau) \dfrac{P^1_{-\frac{1}{2}+i\tau}(-\cos\theta)}{P^1_{-\frac{1}{2}+i\tau}(-\cos\alpha)} e^{-i\tau \ln(r/a)} \, d\tau, & \alpha < \theta < \pi, \end{cases} \tag{5.47}$$

for the boudnary condition (5.44) and in the form

$$u = \begin{cases} \sqrt{\dfrac{a}{r}} \displaystyle\int_{-\infty}^{\infty} g(\tau) \dfrac{\sin\alpha \ P^1_{-\frac{1}{2}+i\tau}(\cos\theta)}{P^2_{-\frac{1}{2}+i\tau}(\cos\alpha)} e^{-i\tau \ln(r/a)} \, d\tau, & 0 \leq \theta < \alpha, \\[4mm] -\sqrt{\dfrac{a}{r}} \displaystyle\int_{-\infty}^{\infty} g(\tau) \dfrac{\sin\alpha \ P^1_{-\frac{1}{2}+i\tau}(-\cos\theta)}{P^2_{-\frac{1}{2}+i\tau}(-\cos\alpha)} e^{-i\tau \ln(r/a)} \, d\tau, & \alpha < \theta < \pi, \end{cases} \tag{5.48}$$

for the boundary condition (5.45). The functions $f(\tau)$ and $g(\tau)$ are unknown continuous functions to be determined. The representations (5.47) and (5.48) satisfy formally the differential equation (5.42), the conditions (5.46) and also the continuity conditions on the axis of symmetry and on the surface $\theta = \alpha$. Then the boundary conditions (5.44) and (5.45) and the continuity requirements of the normal derivatives of the unknown function on $\theta = \alpha$ reduce the above mixed problem to the solution of the dual integral equations for the determination of the functions $f(\tau)$ and $g(\tau)$ in the following form:

$$\sqrt{\frac{a}{r}} \int_{-\infty}^{\infty} f(\tau) \, e^{-i\tau \, \ln(r/a)} \, d\tau = f_1(r), \ 0 \le r < a,$$

$$\sqrt{\frac{a}{r}} \int_{-\infty}^{\infty} \frac{f(\tau) \, (\frac{1}{2} + 2\tau^2) \, \cosh \pi\tau}{\sin \alpha \, P^1_{-\frac{1}{2}+i\tau}(\cos \alpha) \, P^1_{-\frac{1}{2}+i\tau}(-\cos \alpha)} \, e^{-i\tau \, \ln(r/a)} \, d\tau = 0, \ r > a, \qquad (5.49)$$

in the problem when displacement is prescribed, and

$$\sqrt{\frac{a}{r}} \int_{-\infty}^{\infty} g(\tau) \, e^{-i\tau \, \ln(r/a)} d\tau = \frac{f_2(r)}{\mu}, \ 0 \le r < a,$$

$$\sqrt{\frac{a}{r}} \int_{-\infty}^{\infty} g(\tau) \, \frac{(\frac{1}{2} + 2\tau^2) \, \cosh \pi\tau}{\pi \, P^2_{-\frac{1}{2}+i\tau}(\cos \alpha) \, P^2_{-\frac{1}{2}+i\tau}(-\cos \alpha)} \, e^{-i\tau \, \ln(r/a)} d\tau = 0, \ r > a, \qquad (5.50)$$

in the problem when shear stress is prescribed.

These pairs of integral equations (5.49) and (5.50) belong to the class of dual integral equations related to the Mellin integral transform (cf. Zlatina 1972).

The solution of the pair (5.49) can be expressed by

$$f(\tau) = \frac{\sin \alpha \, P^1_{-\frac{1}{2}+i\tau}(\cos \alpha) \, P^1_{-\frac{1}{2}+i\tau}(-\cos \alpha) \, \Gamma(\frac{5}{4} + \frac{i\tau}{2})}{4\sqrt{a\pi} \, (1 + 4\tau^2) \, \cosh \pi\tau \, \Gamma(\frac{3}{4} + \frac{i\tau}{2})}$$

$$\times \int_0^a \frac{\phi(t)}{\sqrt{t}} \, e^{i\tau \, \ln(t/a)} \, dt, \qquad (5.51)$$

where $\phi(t)$ is an unknown function to be determined such that it is continuous together with its derivative in $(0, a)$ and $\sqrt{t} \, \phi(t) \to 0$ as $t \to 0$.

Using the relation

$$\frac{1}{4\sqrt{\pi}} \sqrt{\frac{a}{r}} \int_{-\infty}^{\infty} \frac{\Gamma(\frac{1}{4} + \frac{i\tau}{2})}{\Gamma(\frac{3}{4} + \frac{i\tau}{2})} \, e^{-i\tau \, \ln(r/a)} \, d\tau = \begin{cases} \dfrac{a}{\sqrt{a^2 - r^2}}, & r < a, \\ 0, & r > a, \end{cases}$$

it is seen that the second relation of (5.49) is identically satisfied by the expression (5.51). Then substituting (5.51) into the first relation of (5.49), we obtain a Fredholm integral equation of the second kind as

$$\Phi(x) = p(x) - \int_0^{\infty} q(|x - y|) \, \Phi(y) \, dy, \ 0 < x < \infty, \qquad (5.52)$$

where

$$t = a\,e^{-x}, \; e^{-\frac{x}{2}}\,\phi(ae^{-x}) = \Phi(x),$$

$$p(x) = -16\,e^{\frac{3x}{2}}\,\frac{d}{dx}\int_x^\infty \frac{f_1(ae^{-y})\,e^{-3y}}{\sqrt{e^{-2x}-e^{-2y}}}\,dy,$$

and

$$q(x) = \frac{1}{\pi}\int_0^\infty \left\{ \frac{\sin\alpha\,P^1_{-\frac{1}{2}+i\tau}(\cos\alpha)\,P^1_{-\frac{1}{2}+i\tau}(-\cos\alpha)}{\left[P^1_{-\frac{1}{2}+i\tau}(0)\right]^2} - 1 \right\}\cos\tau x\,d\tau.$$

Similarly, the solution of the pair (5.50) can be represented by

$$g(\tau) = \frac{(\frac{1}{2}-i\tau)\,P^2_{-\frac{1}{2}+i\tau}(\cos\alpha)\,P^2_{-\frac{1}{2}+i\tau}(-\cos\alpha)\,\Gamma(\frac{3}{4}+\frac{i\tau}{2})}{2\sqrt{\pi a}\,(\frac{1}{4}+\tau^2)^2\,\cosh\pi\tau\,\Gamma(\frac{1}{4}+\frac{i\tau}{2})}$$

$$\times \int_0^a \frac{\phi(t)}{\sqrt{t}}\,e^{i\tau\,\ln(t/a)}\,dt. \tag{5.53}$$

Using the relation

$$\frac{1}{4\sqrt{\pi}}\sqrt{\frac{a}{r}}\int_{-\infty}^\infty \frac{\Gamma(\frac{3}{4}-\frac{i\tau}{2})}{\Gamma(\frac{5}{4}-\frac{i\tau}{2})}\,e^{-i\tau\,\ln(r/a)}d\tau = \begin{cases} 0, & r < a, \\ \dfrac{a^2}{r\sqrt{r^2-a^2}}, & r > a, \end{cases}$$

and the expression (5.53), the pair (5.50) reduces to the solution of a Fredholm integral equation

$$\Phi(x) = p(x) - \int_0^\infty q(|x-y|)\,\Phi(y)\,dy,\; 0 < x < \infty, \tag{5.54}$$

where

$$t = a\,e^{-x}, \; e^{-\frac{x}{2}}\,\phi(a\,e^{-x}) = \Phi(x),$$

$$p(x) = -\frac{1}{2\pi}e^{\frac{x}{2}}\int_x^\infty \frac{f_2(ae^{-y})\,e^{-2y}}{\sqrt{e^{-2x}-e^{-2y}}}\,dy,$$

and

$$q(x) = \frac{1}{\pi}\int_0^\infty \left\{ \frac{\sin\alpha\,P^2_{-\frac{1}{2}+i\tau}(\cos\alpha)\,P^2_{-\frac{1}{2}+i\tau}(-\cos\alpha)}{\left[P^2_{-\frac{1}{2}+i\tau}(0)\right]^2} - 1 \right\}\cos\tau x\,d\tau.$$

Now exploiting the expansion formula of the product of associated Legendre functions into series (cf. Zlatina 1972)

$$\sin \alpha \; P_{-\frac{1}{2}+i\tau}^{m}(\cos \alpha) \; P_{-\frac{1}{2}+i\tau}^{m}(-\cos \alpha) = \frac{\cosh^2 \pi\tau}{2^m \; \pi^2} \; \frac{\Gamma^2(\frac{1}{2}+i\tau+m)\Gamma^2(\frac{1}{2}-i\tau+m)}{\Gamma^2(1+m)}$$

$$\times \sum_{k=0}^{\infty} (-1)^k \frac{(m+\frac{1}{2})_k}{k!} {}_3F_2 \left(\frac{1}{2}-k+m, \; \frac{1}{2}+i\tau+m, \; \frac{1}{2}-i\tau+m; \; 2m+1, \; m+1; \; 1 \right)$$

$$\times \cot^{2k} \alpha, \tag{5.55}$$

where ${}_3F_2$ is the generalized hypergeometric function, the kernel $q(x)$ of the integral equation (5.52) or (5.54) can be represented in the form

$$q(x) = \frac{1}{\pi} \int_0^{\infty} \left\{ \frac{\sin \alpha \; P_{-\frac{1}{2}+i\tau}^{m}(\cos \alpha) \; P_{-\frac{1}{2}+i\tau}^{m}(-\cos \alpha)}{\left[P_{-\frac{1}{2}+i\tau}^{m}(0) \right]^2} - 1 \right\} \cos \tau x \; d\tau$$

$$= \sum_{k=1}^{\infty} q_k(x) \cot^{2k} x, \tag{5.56}$$

where

$$q_k(x) = (-1)^k \frac{(m+\frac{1}{2})_k}{k!} \int_0^{\infty} \frac{\omega_k(\tau) \left[P_{-\frac{1}{2}+i\tau}^{m}(0) \right]^2}{2^m \; \pi^2 \; (m!)^2} \cos \tau x \; d\tau,$$

$$\omega_k(\tau) = \frac{\Gamma^2(\frac{1}{2}+i\tau+m) \; \Gamma^2(\frac{1}{2}-i\tau+m)}{\left[P_{-\frac{1}{2}+i\tau}^{m}(0) \right]^2} \cosh^2 \pi\tau$$

$$\times {}_3F_2 \left(\frac{1}{2}-k+m, \; \frac{1}{2}+i\tau+m, \; \frac{1}{2}-i\tau+m; \; 2m+1, \; m+1; \; 1 \right).$$

The above series (5.56) converges rapidly for small values of $\cot^2 \alpha$, that is for α close to $\pi/2$ (cone with large apex angle). The coefficients of the series can be obtained by numerical integration and the values of the functions $\omega_k(\tau)$ are determined by the relations

$$\omega_0 = 2^m, \quad \omega_1 = \frac{8\Gamma^2(\frac{3}{4}+\frac{m}{2}+\frac{i\tau}{2}) \; \Gamma^2(\frac{3}{4}-\frac{m}{2}+\frac{i\tau}{2})}{\Gamma^2(\frac{1}{4}-\frac{m}{2}+\frac{i\tau}{2}) \; \Gamma^2(\frac{1}{4}+\frac{m}{2}+\frac{i\tau}{2})} + \frac{1}{2} - 2\tau^2 - 2m^2,$$

and

$$\left[\left(k+\frac{m}{2}+\frac{1}{2}\right)^2+\frac{m^2}{4}\right]\omega_{k+1}+\left(\tau^2-\frac{1}{4}-2k^2\right)\omega_k+\left[\left(k-\frac{m}{2}-\frac{1}{2}\right)^2+\frac{m^2}{4}\right]\omega_{k-1}=0.$$

The expansion of the kernel $q(x)$ of (5.52) in the problem with prescribed displacements and of (5.54) in the problem with prescribed shear stresses can be obtained from (5.56) as particular cases by putting $m=1$ and $m=2$ respectively.

Thus, the solution of the Fredholm integral equation (5.52) or (5.54) can be obtained in the form

$$\Phi(x)=\sum_{k=0}^{\infty}\Phi_k(x)\ \cot^{2k}\alpha, \tag{5.57}$$

where

$$\Phi_0(x) = p(x),$$
$$\Phi_k(x) = -\sum_{n=1}^{k}\int_0^{\infty}q_n(|x-y|)\ \Phi_{k-n}(y)\ dy,\ (k=1,2,3,\ldots).$$

The method discussed above, to obtain the solution of a class of dual integral equations connected with the Mellin transform, can be applied to find the solution of a physical problem of the torsion of the elastic space with an inclusion in the form of a rigid thin cone. In this problem, the tangential displacement can be expressed by the sum of displacements originating from the torsion of a homogeneous space and additional displacements induced by the presence of the cone as given by

$$u_\phi = \frac{1}{2}\ \gamma\ r^2\ \sin 2\theta + u$$

where γ is the constant angle of twist per unit length. Then this problem reduces to the problem discussed above for

$$f_1(r) = \theta r - \frac{1}{2}\ \gamma\ r^2\ \sin 2\alpha,$$

where θ is the unknown angle of rotation of the cone to be determined. The value of this angle can be obtained from the condition that the moment of the shear stress acting upon the cone is zero.

185

Applying the method developed above, the problem reduces to the solution of the Fredholm integral equation of the second kind as given by

$$\Phi(x) = 16\theta \, e^{-\frac{3x}{2}} + 3\pi\gamma \, a \sin 2\alpha \, e^{-\frac{5x}{2}} - \int_0^\infty q(|x-y|) \, \Phi(y) \, dy, \quad 0 \le x < \infty, \quad (5.58)$$

where the kernel $q(x)$ is defined by (5.56) for $m = 1$. The solution of this integral equation can be represented by

$$\Phi(x) = 64\theta \, \Phi_1(x) - 15\pi\gamma \, a \sin 2\alpha \, \Phi_2(x),$$

where $\Phi_1(x)$ and $\Phi_2(x)$ are the solutions of the integral equations

$$\Phi_1(x) = 4e^{-\frac{3x}{2}} - \int_0^\infty q(|x-y|) \, \Phi_1(y) \, dy$$

and

$$\Phi_2(x) = 5e^{-\frac{5x}{2}} - \int_0^\infty q(|x-y|) \, \Phi_2(y) \, dy$$

respectively. The functions $\Phi_1(x)$ and $\Phi_2(x)$ can be expressed in the form of the series

$$\Phi_1(x) = \sum_{k=0}^\infty \Phi_{k,1}(x) \, \cot^{2k} \alpha,$$

$$\Phi_2(x) = \sum_{k=0}^\infty \Phi_{k,2}(x) \, \cot^{2k} \alpha,$$

where the coefficients of these series can be computed by the scheme mentioned above. After determining the functions $\Phi_1(x)$ and $\Phi_2(x)$, the resulting moment applied to the cone is given by the relation

$$M = 3\pi^2 \, a^2\mu \, \sin^2\alpha \int_0^\infty [64\theta \, \Phi_1(x) - 15\pi\gamma \, a \sin 2\alpha \, \Phi_2(x)] \, e^{-\frac{5x}{2}} \, dx.$$

Now, the condition $M = 0$ gives the relation between the angle θ and the given twisting angle γ as

$$\frac{\theta k}{\gamma} = \left(\int_0^\infty \Phi_2(x) \, e^{-\frac{5x}{2}} dx\right) \Big/ \left(\int_0^\infty \Phi_1(x) \, e^{-\frac{5x}{2}} dx\right),$$

where $k = 64 \, (15\pi a \, \sin 2\alpha)^{-1}$.

For the values of α which are close to $\pi/2$, a simple expression in the form of an expression in powers of $\cot^2\alpha$ is obtained as

$$\frac{\theta k}{\gamma} = 1 + 0.3578 \, \cot^2\alpha + 0.0976 \, \cot^4\alpha + \dots$$

Chapter 6

Hybrid dual integral equations

The dual integral equations presented in the previous chapter arise usually in the study of some mixed boundary value problems in mathematical physics involving homogeneous media. In these dual integral equations the same special function appear in the kernels of both the integral equations. Virchenko (1984) mentioned that Protsenko and Solov'ev (1982), while considering mixed boundary value problems involving piecewise non-homogeneous media, obtained dual integral equations with principally different kernels, and called these hybrid dual integral equations.

In this chapter, we present methods as well as solutions of a number of certain hybrid dual integral equations with mixed kernels considered by Virchenko (1984) and some other equations in two sections. These pairs of integral equations are not widely known in the literature.

6.1 Mixed kernels with generalized associated Legendre functions of the first kind and trigonometric functions

In this section, we consider some hybrid dual integral equations considered by Virchenko (1984) in which one equation contains the generalized associated Legendre function of the first kind as its kernel while the other equation contains trignometric function as kernels. However, the specific physical problems in which these hybrid dual integral equations occur were however not mentioned by Virchenko (1984). Four

187

pairs of integral equations are considered here. The first two pairs are solved explicitly by exploiting integral representations of the generalized associated Legendre function of the first kind. The last two pairs contain a weight function and as such these are reduced to Fredholm integral equations of the second kind.

First we consider the dual integral equations

$$\int_0^\infty f(\tau) \, P_{-\frac{1}{2}+i\tau}^{\mu,\nu}(\cosh x) \, d\tau \;=\; g_1(x), \; 0 \le x \le a, \tag{6.1}$$

$$\int_0^\infty f(\tau) \, \cos x\tau \, d\tau \;=\; g_2(x), \; a < x < \infty, \tag{6.2}$$

where $f(\tau)$ is the unknown function, $g_1(x)$ and $g_2(x)$ are known functions and $P_{-\frac{1}{2}+i\tau}^{\mu,\nu}(z)$ is the generalized associated Legendre function of the first kind.

The following two integral representations (see Appendix A-3)

$$P_{-\frac{1}{2}+i\tau}^{\mu,\nu}(\cosh x) \;=\; \frac{2^{\frac{\nu-\mu+1}{2}} \sinh^\mu x}{\sqrt{\pi} \, \Gamma(\frac{1}{2}-\mu)} \int_0^x \frac{\cos \alpha\tau}{(\cosh x - \cosh \alpha)^{\frac{1}{2}+\mu}}$$
$$\times F\left(\frac{\nu-\mu}{2}, -\frac{\nu+\mu}{2}; \frac{1}{2}-\mu; \frac{\cosh x - \cosh \alpha}{1+\cosh x}\right) d\alpha, \tag{6.3}$$

and

$$\cos \alpha\tau \;=\; \frac{\Gamma\left(\frac{1}{2}-\mu\right) \cos \pi\mu}{\sqrt{\pi} \, 2^{\frac{\nu-\mu+1}{2}}} \frac{d}{d\alpha} \left[(\cosh \alpha + 1)^{\frac{\nu-\mu}{2}} \int_0^\alpha (\cosh \alpha - \cosh x)^{\mu-\frac{1}{2}}\right.$$
$$\times (\cosh x + 1)^{\frac{\mu-\nu}{2}} F\left(\frac{\mu-\nu}{2}, \frac{1+\mu-\nu}{2}; \frac{1}{2}+\mu; \frac{\cosh \alpha - \cosh x}{1+\cosh \alpha}\right)$$
$$\left. \times P_{-\frac{1}{2}+i\tau}^{\mu,\nu}(\cosh x) \, \sinh^{1-\mu} x \, dx\right], \tag{6.4}$$

have been used in the process of obtaining the solution of the above pair of hybrid integral equations.

Now multiplying the relation (6.1) by the expression given by

$$\frac{2^{\frac{\mu-\nu}{2}}}{\pi} \Gamma\left(\frac{1}{2}-\mu\right) \cos \pi\mu \, \sinh^{1-\mu} x \, (\cosh \alpha - \cosh x)^{\mu-\frac{1}{2}} (\cosh \alpha + 1)^{\frac{\nu-\mu}{2}}$$

$$\times (\cosh x + 1)^{\frac{\mu-\nu}{2}} F\left(\frac{\mu-\nu}{2}, \frac{1+\mu-\nu}{2}; \frac{1}{2}+\mu; \frac{\cosh \alpha - \cosh x}{1+\cosh \alpha}\right), \tag{6.5}$$

and integrating with respect to x from 0 to α and then differentiating with respect to α and interchanging the order of integration, we get

$$\frac{\Gamma\left(\frac{1}{2}-\mu\right)\cos\pi\mu}{\pi\,2^{\frac{\nu-\mu}{2}}}\int_0^\infty f(\tau)\,\frac{d}{d\alpha}\left[(\cosh\alpha+1)^{\frac{\nu-\mu}{2}}\int_0^\alpha\frac{\sinh^{1-\mu}x\;P^{\mu,\nu}_{-\frac{1}{2}+i\tau}(\cosh x)}{(\cosh\alpha-\cosh x)^{\frac{1}{2}-\mu}}\right.$$

$$\left.\times(\cosh x+1)^{\frac{\mu-\nu}{2}}\,F\left(\frac{\mu-\nu}{2},\,\frac{1+\mu-\nu}{2};\,\frac{1}{2}+\mu;\,\frac{\cosh\alpha-\cosh x}{1+\cosh\alpha}\right)\,dx\right]d\tau$$

$$=G_1(\alpha),\;0\le\alpha\le a,\tag{6.6}$$

where

$$\begin{aligned}
G_1(\alpha) &= \frac{2^{\frac{\mu-\nu}{2}}\Gamma\left(\frac{1}{2}-\mu\right)}{\pi}\cos\pi\mu\,\frac{d}{d\alpha}\left[(\cosh\alpha+1)^{\frac{\nu-\mu}{2}}\right.\\
&\quad\times\int_0^\alpha\frac{\sinh^{1-\mu}x\;g_1(x)}{(\cosh\alpha-\cosh x)^{\frac{1}{2}-\mu}}\,(\cosh x+1)^{\frac{\mu-\nu}{2}}\\
&\quad\left.\times F\left(\frac{\mu-\nu}{2},\,\frac{1+\mu-\nu}{2};\,\frac{1}{2}+\mu;\,\frac{\cosh\alpha-\cosh x}{1+\cosh\alpha}\right)\,dx\right].
\end{aligned}\tag{6.7}$$

Using the relation (6.4), equation (6.6) can be written as

$$F_c\left[f(\tau);\alpha\right]=G_1(\alpha),\;0\le\alpha\le a,\tag{6.8}$$

where $F_c[f(\tau);\alpha]$ is the Fourier cosine transform of $f(\tau)$. Equation (6.2) can be rewritten as

$$F_c\left[f(\tau);\alpha\right]=G_2(\alpha),\;a<\alpha<\infty,\tag{6.9}$$

where $G_2(\alpha)=\sqrt{\frac{2}{\pi}}\,g_2(\alpha)$.

Hence, by the Fourier cosine inversion theorem, the relations (6.8) and (6.9) produce

$$f(\tau)=\sqrt{\frac{2}{\pi}}\left[\int_0^a G_1(\alpha)\,\cos\alpha\tau\,d\alpha+\int_a^\infty G_2(\alpha)\,\cos\alpha\tau\,d\alpha\right],\tag{6.10}$$

which gives the solution of the hybrid pair of integral equations (6.1) and (6.2).

Now, we consider the mixed pair of integral equations

$$\int_0^\infty \tau\,f(\tau)\,P^{\mu,\nu}_{-\frac{1}{2}+i\tau}(\cosh x)\,d\tau = g_3(x),\;0\le x\le a,\tag{6.11}$$

$$\int_0^\infty f(\tau)\,\sin x\tau\,d\tau = g_4(x),\;a<x<\infty,\tag{6.12}$$

189

where $g_3(x)$ and $g_4(x)$ are known functions.

Multiplying the relation (6.11) on both sides by the expression (6.5) and integrating with respect to x from 0 to α and then interchanging the order of integration, we obtain

$$
\frac{2^{\frac{\mu-\nu}{2}}}{\pi} \Gamma\left(\frac{1}{2}-\mu\right) \cos\pi\mu \int_0^\infty \tau\, f(\tau)\,(\cosh\alpha+1)^{\frac{\nu-\mu}{2}} \left[\int_0^\alpha (\cosh\alpha-\cosh x)^{\mu-\frac{1}{2}}\right.
$$
$$
\times \sinh^{1-\mu} x\, F\left(\frac{\mu-\nu}{2},\frac{1+\mu-\nu}{2};\frac{1}{2}+\mu;\frac{\cosh\alpha-\cosh x}{1+\cosh\alpha}\right) P^{\mu,\nu}_{-\frac{1}{2}+i\tau}(\cosh x)\, dx
$$
$$
= G_3(\alpha),\ 0\le\alpha\le a, \tag{6.13}
$$

where

$$
G_3(\alpha) = \frac{2^{\frac{\mu-\nu}{2}}}{\pi}\Gamma\left(\frac{1}{2}-\mu\right)\cos\pi\mu\,(\cosh\alpha+1)^{\frac{\nu-\mu}{2}}\int_0^\alpha(\cosh\alpha-\cosh x)^{\mu-\frac{1}{2}}\ \sinh^{1-\mu} x
$$
$$
\times(\cosh x+1)^{\frac{\mu-\nu}{2}} F\left(\frac{\mu-\nu}{2},\frac{1+\mu-\nu}{2};\frac{1}{2}+\mu;\frac{\cosh\alpha-\cosh x}{1+\cosh\alpha}\right)
$$
$$
\times g_3(x)\, dx. \tag{6.14}
$$

Using the relation (see Appendix A-3, Eq. (A.3.12))

$$
\sin\alpha\tau = \frac{2^{\frac{\mu-\nu-1}{2}}}{\sqrt{\pi}}\Gamma\left(\frac{1}{2}-\mu\right)\tau\cos\pi\mu\,(\cosh\alpha+1)^{\frac{\nu-\mu}{2}}\int_0^\alpha(\cosh\alpha-\cosh x)^{\mu-\frac{1}{2}}
$$
$$
\times(\cosh x+1)^{\frac{\mu-\nu}{2}} F\left(\frac{\mu-\nu}{2},\frac{1+\mu-\nu}{2};\frac{1}{2}+\mu;\frac{\cosh\alpha-\cosh x}{1+\cosh\alpha}\right)
$$
$$
\times P^{\mu,\nu}_{-\frac{1}{2}+i\tau}(\cosh x)\ \sinh^{1-\mu} x\, dx,
$$

the above equation (6.13) can be reduced to

$$
F_s[f(\tau);\ \alpha] = G_3(\alpha),\ 0\le\alpha\le a, \tag{6.15}
$$

where F_s denotes the Fourier sine transform.

Equation (6.12) can be rewritten as

$$
F_s[f(\tau);\alpha] = G_4(\alpha),\ a<\alpha<\infty, \tag{6.16}
$$

where $G_4(\alpha) = \sqrt{\frac{2}{\pi}}\,g_4(\alpha)$.

Hence, by the Fourier sine inversion theorem, the relations (6.15) and (6.16) produce the solution of the mixed pair of integral equations (6.11) and (6.12) as

$$f(\tau) = \sqrt{\frac{2}{\pi}} \left[\int_0^a G_3(\alpha) \, \sin \alpha \tau \, d\alpha + \int_a^\infty G_4(\alpha) \, \sin \alpha \tau \, d\alpha \right].$$

Next, we consider the hybrid pair of integral equations of the more general form

$$\int_0^\infty f(\tau) \left[1 + \omega(\tau) \right] P_{-\frac{1}{2}+i\tau}^{\mu,\nu}(\cosh x) \, d\tau \;=\; g_5(x), \; 0 \le x \le a, \qquad (6.17)$$

$$\int_0^\infty f(\tau) \cos x\tau \, d\tau \;=\; g_2(x), \; a < x < \infty, \qquad (6.18)$$

where $\omega(\tau)$, $g_5(x)$ and $g_2(x)$ are known functions.

The solution of the above pair of integral equations can be reduced to that of a Fredholm integral equation of the second kind.

The relation (6.17) can be rewritten as

$$\int_0^\infty f(\tau) \, P_{-\frac{1}{2}+i\tau}^{\mu,\nu}(\cosh x) \, d\tau = g_5(x) - \int_0^\infty f(\tau) \, \omega(\tau) \, P_{-\frac{1}{2}+i\tau}^{\mu,\nu}(\cosh x) \, d\tau, \; 0 \le x \le a.$$
$$(6.19)$$

The dual integral equations (6.19) and (6.18) are of the form (6.1) and (6.2) respectively. Therefore, using the relation (6.10), the solution of this pair of integral equations can be obtained as

$$f(\tau) = \sqrt{\frac{2}{\pi}} \left[\int_0^a G_5(\alpha) \, \cos \alpha \tau \, d\alpha + \int_a^\infty G_2(\alpha) \, \cos \alpha \tau \, d\alpha \right], \qquad (6.20)$$

where

$$G_2(\alpha) \;=\; \sqrt{\frac{2}{\pi}} \, g_2(\alpha), \qquad (6.21)$$

$$G_5(\alpha) \;=\; \frac{2^{\frac{\mu-\nu}{2}}\Gamma(\frac{1}{2}-\mu)}{\pi} \, \cos \pi\mu \, \frac{d}{d\alpha} \left[(\cosh \alpha + 1)^{\frac{\nu-\mu}{2}} \int_0^\alpha \frac{\sinh^{1-\mu} x}{(\cosh \alpha - \cosh x)^{\frac{1}{2}-\mu}} \right.$$

$$\times (\cosh x + 1)^{\frac{\mu-\nu}{2}} F\left(\frac{\mu-\nu}{2}, \frac{1+\mu-\nu}{2}; \frac{1}{2}+\mu; \frac{\cosh \alpha - \cosh x}{1+\cosh \alpha} \right)$$

$$\times \left. \left\{ g_5(x) - \int_0^\infty f(\tau) \, \omega(\tau) P_{-\frac{1}{2}+i\tau}^{\mu,\nu}(\cosh x) \, d\tau \right\} dx \right]. \qquad (6.22)$$

Let us define

$$\overline{G}_5(\alpha) = \frac{2^{\frac{\mu-\nu}{2}}\Gamma(\frac{1}{2}-\mu)}{\pi} \cos\pi\mu \frac{d}{d\alpha}\left[(\cosh\alpha+1)^{\frac{\nu-\mu}{2}}\int_0^\alpha \frac{\sinh^{1-\mu}x}{(\cosh\alpha-\cosh x)^{\frac{1}{2}-\mu}}\right.$$

$$\times(\cosh x+1)^{\frac{\mu-\nu}{2}}F\left(\frac{\mu-\nu}{2},\frac{1+\mu-\nu}{2};\frac{1}{2}+\mu;\frac{\cosh\alpha-\cosh x}{1+\cosh\alpha}\right)$$

$$\times g_5(x)\ dx] \tag{6.23}$$

and

$$H(\tau) = \sqrt{\frac{2}{\pi}}\int_a^\infty G_2(\alpha)\ \cos\alpha\tau\ d\alpha. \tag{6.24}$$

Then, by the relation (6.4),

$$\frac{2^{\frac{\mu-\nu}{2}}\Gamma(\frac{1}{2}-\mu)}{\pi} \cos\pi\mu \frac{d}{d\alpha}\left[(\cosh\alpha+1)^{\frac{\nu-\mu}{2}}\int_0^\alpha \frac{\sinh^{1-\mu}x}{(\cosh\alpha-\cosh x)^{\frac{1}{2}-\mu}}\right.$$

$$\times(\cosh x+1)^{\frac{\mu-\nu}{2}}F\left(\frac{\mu-\nu}{2},\frac{1+\mu-\nu}{2};\frac{1}{2}+\mu;\frac{\cosh\alpha-\cosh x}{1+\cosh\alpha}\right)$$

$$\times\left\{\int_0^\infty f(\tau)\ \omega(\tau)\ P^{\mu,\nu}_{-\frac{1}{2}+i\tau}(\cosh x)\ d\tau\right\}dx]$$

$$= \sqrt{\frac{2}{\pi}}\int_0^\infty f(\tau)\ \omega(\tau)\ \cos\alpha\tau\ d\tau$$

$$= \frac{2}{\pi}\int_0^a G_5(\beta)\left\{\int_0^\infty \omega(\tau)\ \cos\alpha\tau\ \cos\beta\tau\ d\tau\right\}d\beta$$

$$+\sqrt{\frac{2}{\pi}}\int_0^\infty H(\tau)\ \omega(\tau)\ \cos\alpha\tau\ d\tau.$$

Hence, the solution of the hybrid pair of integral equations (6.17) and (6.18) is given by (6.20), where $G_5(\alpha)$ is the solution of the Fredholm integral equation of the second kind

$$G_5(\alpha) + \int_0^a G_5(\beta)\ K(\alpha,\beta)\ d\beta = G^*(\alpha),\ 0\le x\le a, \tag{6.25}$$

192

with the kernel $K(\alpha, \beta)$ defined by

$$
\begin{aligned}
K(\alpha, \beta) &= \frac{1}{\sqrt{2\pi}} \left\{ \omega_c(\alpha + \beta) + \omega_c(|\alpha - \beta|) \right\}, \\
\omega_c(\alpha) &= F_c\left[\omega(\tau); \alpha\right],
\end{aligned}
$$

and

$$
\begin{aligned}
G^*(\alpha) &= \overline{G}_5(\alpha) - \overline{\overline{G}}_5(\alpha), \\
\overline{\overline{G}}_5(\alpha) &= F_c\left[H(\tau)\,\omega(\tau); \alpha\right].
\end{aligned}
$$

Similarly, the solution of the general pair of mixed integral equations

$$
\int_0^\infty \tau\, f(\tau)\, P_{-\frac{1}{2}+i\tau}^{\mu,\nu}(\cosh x)\, d\tau = g_3(x),\ 0 \le x \le a, \tag{6.26}
$$

$$
\int_0^\infty f(\tau)\, [1 + \omega(\tau)] \sin x\tau\, d\tau = g_6(x),\ a < x < \infty, \tag{6.27}
$$

is given by

$$
f(\tau) = \sqrt{\frac{2}{\pi}} \left[\int_0^a G_3(\alpha)\, \sin \alpha\tau\, d\alpha + \int_a^\infty G_6(\alpha)\, \sin \alpha\tau\, d\alpha \right], \tag{6.28}
$$

where $G_3(\alpha)$ is given by (6.14) and $G_6(\alpha)$ is the solution of the Fredholm integral equation of the second kind

$$
G_6(\alpha) = H^*(\alpha) + \int_a^\infty G_6(\beta)\, K_1(\alpha, \beta)\, d\beta,\ a < \alpha < \infty, \tag{6.29}
$$

where the kernel $K_1(\alpha, \beta)$ is defined by

$$
\begin{aligned}
K_1(\alpha, \beta) &= \frac{1}{\sqrt{2\pi}} \left\{ \omega_c(\alpha + \beta) - \omega_c(|\alpha - \beta|) \right\}, \\
\omega_c(\alpha) &= F_c\left[\omega(\tau); \alpha\right],
\end{aligned}
$$

and the function $H^*(\alpha)$ is defined by

$$
\begin{aligned}
H^*(\alpha) &= H_1(\alpha) - H_2(\alpha), \\
H_1(\alpha) &= \sqrt{\frac{2}{\pi}}\, g_6(\alpha),
\end{aligned}
$$

$$
H_2(\alpha) = \frac{2}{\pi} \int_0^a G_3(x) \left\{ \int_0^a \omega(\tau)\, \sin \alpha\tau\, \sin x\tau\, d\tau \right\} dx.
$$

6.2 Mixed kernels involving Bessel functions of the first kind of order zero and trigonometric functions

This section is concerned with some hybrid dual integral equations wherein one of the equations involves the Bessel function of the first kind of order zero as kernel and the other equation involves a trigonometric function as kernel. Four pairs of mixed integral equations are considered here. The first two pairs are solved in closed form by using integral representations of the Bessel function in terms of trigonometric functions. The last two pairs contain a weight function and as such these are reduced to the Fredholm integral equations of the second kind where solutions can be obtained at least numerically.

First we consider the mixed pair of integral equations

$$\int_0^\infty f(\tau) \, J_0(x\tau) \, d\tau = g(x), \ 0 \le x \le a, \tag{6.30}$$

$$\int_0^\infty f(\tau) \, \cos x\tau \, d\tau = h(x), \ a < x < \infty, \tag{6.31}$$

where $g(x)$, $h(x)$ are known functions. To obtain the solution of these hybrid dual integral equations, multiplying (6.30) by $(2/\pi)^{1/2} x(\xi^2 - x^2)^{-1/2}$ and integrating with respect to x from 0 to ξ ($\le a$), and then differentiating with respect to ξ, and interchanging the order of integration and using the relation

$$\cos \xi\tau = \frac{d}{d\xi} \int_0^\xi \frac{x \, J_0(x\tau)}{\sqrt{\xi^2 - x^2}} \, dx, \ \tau > 0, \ \xi > 0, \tag{6.32}$$

it becomes

$$\sqrt{\frac{2}{\pi}} \int_0^\infty f(\tau) \, \cos \xi\tau \, d\tau = g_1(\xi), \ 0 \le \xi \le a, \tag{6.33}$$

where

$$g_1(\xi) = \sqrt{\frac{2}{\pi}} \frac{d}{d\xi} \int_0^\xi \frac{x \, g(x)}{\sqrt{\xi^2 - x^2}} \, dx. \tag{6.34}$$

Let $F_c(\xi)$ denote the Fourier cosine transform of $f(\tau)$. Then (6.33) can be represented as

$$F_c[f(\tau); \xi] = g_1(\xi), \ 0 \le \xi \le a. \tag{6.35}$$

Also the equation (6.31) can be written as

$$F_c[f(\tau); \xi] = h_1(\xi); \ a < \xi < \infty, \tag{6.36}$$

where

$$h_1(\xi) = \sqrt{\frac{2}{\pi}} \ h(\xi). \tag{6.37}$$

Hence, using the inversion formula for the Fourier cosine transform (6.35) and (6.36) produces

$$f(\tau) = \sqrt{\frac{2}{\pi}} \left[\int_0^a g_1(\xi) \ \cos \xi\tau \ d\xi + \int_a^\infty h_1(\xi) \ \cos \xi\tau \ d\xi \right], \tag{6.38}$$

which gives the solution of the hybrid pair (6.30) and (6.31).

Next, the second pair of mixed integral equations is

$$\int_0^\infty \tau \ f(\tau) \ J_0(x\tau) \ d\tau = m(x), \ 0 \le x \le a, \tag{6.39}$$

$$\int_0^\infty f(\tau) \ \sin x\tau \ d\tau = n(x), \ a < x < \infty, \tag{6.40}$$

where $m(x)$ and $n(x)$ are known functions.

Multiplying (6.39) by $(2/\pi)^{1/2} x(\xi^2 - x^2)^{-1/2}$, integrating with respect to x from 0 to ξ ($\le a$) and interchanging the order of integrations and then using the relation

$$\sin \xi\tau = \tau \int_0^\xi \frac{x \ J_0(x\tau)}{\sqrt{\xi^2 - x^2}} \ dx, \ \tau > 0, \ \xi > 0,$$

we obtain

$$\sqrt{\frac{2}{\pi}} \int_0^\infty f(\tau) \ \sin \xi\tau \ d\tau = m_1(\xi), \ 0 \le \xi \le a, \tag{6.41}$$

where

$$m_1(\xi) = \sqrt{\frac{2}{\pi}} \int_0^\xi \frac{x \ m(x)}{\sqrt{\xi^2 - x^2}} \ dx. \tag{6.42}$$

Let $F_s(\xi)$ denote the Fourier sine transform of $f(\tau)$. Then the equation (6.41) becomes

$$F_s[f(\tau); \xi] = m_1(\xi), \ 0 \le \xi \le a. \tag{6.43}$$

195

Also the equation (6.40) is rewritten as

$$F_s[f(\tau); \xi] = n_1(\xi), \ a < \xi < \infty, \tag{6.44}$$

where

$$n_1(\xi) = \sqrt{\frac{2}{\pi}} \, n(\xi). \tag{6.45}$$

Then, the solution of the pair (6.39) and (6.40) is given by

$$f(\tau) = \sqrt{\frac{2}{\pi}} \left[\int_0^a m_1(\xi) \, \sin \xi\tau \, d\xi + \int_a^\infty n_1(\xi) \, \sin \xi\tau \, d\xi \right]. \tag{6.46}$$

Now, we consider the following mixed pair of integral equations of more general form as

$$\int_0^\infty f(\tau) \{1 + \omega(\tau)\} J_0(x\tau) \, d\tau = p(x), \ 0 \le x \le a, \tag{6.47}$$

$$\int_0^\infty f(\tau) \cos x\tau \, d\tau = h(x), \ a < x < \infty, \tag{6.48}$$

where the function $\omega(\tau)$ is a known weight function. Equation (6.47) can be written as

$$\int_0^\infty f(\tau) \, J_0(x\tau) \, d\tau = p(x) - \int_0^\infty f(\tau) \, \omega(\tau) \, J_0(x\tau) \, d\tau, \ 0 \le x \le a. \tag{6.49}$$

Then by the relation (6.38), the solution of the hybrid dual integral equations (6.49) and (6.48) is given by

$$f(\tau) = \sqrt{\frac{2}{\pi}} \left[\int_0^a p_1(\xi) \, \cos \xi\tau \, d\xi + \int_a^\infty h_1(\xi) \, \cos \xi\tau \, d\xi \right], \tag{6.50}$$

where $h_1(\xi)$ is the same as that defined by (6.37) and $p_1(\xi)$ is given by

$$p_1(\xi) = \sqrt{\frac{2}{\pi}} \frac{d}{d\xi} \int_0^\xi x \, (\xi^2 - x^2)^{-1/2} \left\{ p(x) - \int_0^\infty f(\tau) \, \omega(\tau) \, J_0(x\tau) \, d\tau \right\} dx. \tag{6.51}$$

Introducing

$$\overline{p}_1(\xi) = \sqrt{\frac{2}{\pi}} \frac{d}{d\xi} \int_0^\xi \frac{x \, p(x)}{\sqrt{\xi^2 - x^2}} \, dx$$

and

$$H_1(\tau) = \sqrt{\frac{2}{\pi}} \int_a^\infty h_1(\xi) \, \cos \xi\tau \, d\xi,$$

the second integral in (6.51) can be simplified. Specifically,

$$\sqrt{\frac{2}{\pi}} \frac{d}{d\xi} \int_0^\xi \frac{x \, dx}{\sqrt{\xi^2 - x^2}} \int_0^\infty f(\tau) \, \omega(\tau) \, J_0(x\tau) \, d\tau = \sqrt{\frac{2}{\pi}} \int_0^\infty f(\tau) \, \omega(\tau) \, \cos \xi\tau \, d\tau$$

$$= \frac{2}{\pi} \int_0^\infty \omega(\tau) \left[\int_0^a p_1(y) \, \cos y\tau \, dy + \int_a^\infty h_1(y) \, \cos y\tau \, dy \right] \cos \xi\tau \, d\tau$$

$$= \frac{2}{\pi} \int_0^a p_1(y) \, dy \int_0^\infty \omega(\tau) \, \cos \xi\tau \, \cos y\tau \, d\tau + \sqrt{\frac{2}{\pi}} \int_0^\infty \omega(\tau) \, H_1(\tau) \, \cos \xi\tau \, d\tau.$$

Therefore, $p_1(\xi)$ satisfies the Fredholm integral equation of the second kind as given by

$$p_1(\xi) + \int_0^a p_1(u) \, K(\xi, u) \, du = p^*(\xi), \ 0 \le \xi \le a, \tag{6.52}$$

where

$$K(\xi, u) = (2\pi)^{-1/2} \left\{ \omega_c(\xi + u) + \omega_c(|\xi - u|) \right\}$$

with

$$\omega_c(\xi) = F_c \left[\omega(\tau); \ \xi \right],$$

and

$$p^*(\xi) = \overline{p}_1(\xi) - p_2(\xi)$$

with

$$p_2(\xi) = F_c \left[\omega(\tau) H_1(\tau); \xi \right].$$

Similarly, the solution of the mixed pair

$$\int_0^\infty \tau \, f(\tau) \, J_0(x\tau) \, d\tau = m(x), \ 0 \le x \le a,$$

$$\int_0^\infty f(\tau) \left\{ 1 + \omega(\tau) \right\} \sin x\tau \, d\tau = v(x), \ a < x < \infty,$$

is given by

$$f(\tau) = \sqrt{\frac{2}{\pi}} \left[\int_0^a m_1(\xi) \, \sin \xi\tau \, d\xi + \int_a^\infty v_1(\xi) \, \sin \xi\tau \, d\xi \right],$$

where $m_1(\xi)$ is the same as that defined by (6.42), but $v_1(\xi)$ satisfies the Fredholm integral equation of the second kind as given by

$$v_1(\xi) = v^*(\xi) + \int_a^\infty v_1(u) \, K_1(\xi, u) \, du, \ a < \xi < \infty,$$

where

$$K_1(\xi, u) = (2\pi)^{-1/2} \left\{ \omega_c(\xi + u) - \omega_c(|\xi - u|) \right\},$$

$$v^*(\xi) = v_1(\xi) - v_2(\xi),$$

$$v_1(\xi) = (2\pi)^{1/2} v(\xi),$$

and

$$v_2(\xi) = \frac{2}{\pi} \int_0^a m_1(y) \left\{ \int_0^\infty \omega(\tau) \, \sin \xi\tau \, \sin y\tau \, d\tau \right\} dy.$$

Appendix

Useful results of some special functions

Here some useful results on Bessel, Legendre, associated Legendre and generalized associated Legendre functions which have been utilized, are given. The appropriate references for these results are also mentioned.

A-1 Bessel functions

A.1.1 Some integrals involving Bessel functions

The Bessel function of the first kind $J_\nu(x)$ and of the second kind $Y_\nu(x)$ are two independent solutions of the ordinary differential equation

$$x^2 \frac{d^2 u}{dx^2} + x \frac{du}{dx} + (x^2 - \nu^2)u = 0.$$

The following results involving Bessel functions have been utilized in this book (cf. Gradshteyn and Ryzhik 1992).

$$\int_0^\infty \cos xu \ J_0(ut) \ du = \begin{cases} \dfrac{1}{\sqrt{t^2 - x^2}}, & x < t, \\ 0, & x > t, \end{cases} \tag{A.1.1}$$

$$\int_0^\infty \sin xu \ J_0(ut) \ du = \begin{cases} 0, & x < t, \\ \dfrac{1}{\sqrt{x^2 - t^2}}, & x > t, \end{cases} \tag{A.1.2}$$

$$\int_0^\infty \cos xu \ J_1(ut) \ du = \begin{cases} \dfrac{1}{t}, & x < t, \\ \dfrac{-t}{\sqrt{x^2 - t^2}(x + \sqrt{x^2 - t^2})}, & x > t, \end{cases} \tag{A.1.3}$$

$$\int_0^1 u \ J_0(xu) \ du = \frac{J_1(x)}{x}, \tag{A.1.4}$$

$$\int_t^\infty \frac{u \ J_0(xu)}{\sqrt{x^2 - t^2}} \ du = \frac{\cos xt}{x}, \tag{A.1.5}$$

199

$$\int_0^t \frac{J_1(xu)}{\sqrt{t^2 - x^2}}\, du = \frac{2\sin^2\left(\frac{ut}{2}\right)}{ut}, \tag{A.1.6}$$

$$\int_t^\infty \frac{J_1(xu)}{\sqrt{u^2 - t^2}}\, du = \frac{\sin xt}{xt}, \tag{A.1.7}$$

$$\int_0^t \frac{\cos xu}{\sqrt{t^2 - u^2}}\, du = \frac{\pi}{2}\, J_0(xt), \tag{A.1.8}$$

$$\int_0^t \frac{u\sin xu}{\sqrt{t^2 - u^2}}\, du = \frac{\pi}{2} t\, J_1(xt), \tag{A.1.9}$$

$$\frac{\pi}{2}\, J_1(ax) = \int_0^a \frac{\cos xu}{a}\, du$$

$$\tag{A.1.10}$$

$$-\frac{1}{a}\int_a^\infty \left(\frac{u}{\sqrt{u^2 - a^2}} - 1\right)\cos xu\, du, \quad a > 0.$$

$$\int_0^\infty \frac{\sin ut\, J_0(au)}{u^2 + K^2}\, du = \frac{\sinh Kt}{K}\, K_0(Ka), \quad K > 0,\ 0 < t < a, \tag{A.1.11}$$

$$\int_0^\infty \frac{u\sin ut\, J_0(au)}{u^2 + K^2}\, du = \frac{\pi}{2}\, e^{-Kt}\, I_0(Ka), \quad K > 0,\ a < t < \infty, \tag{A.1.12}$$

$$\int_0^\infty \frac{\sin ut\, J_1(bu)}{u^2 + K^2}\, du = \frac{\pi}{2K}\, e^{-Ky}\, I_1(Kb), \quad K > 0,\ b < t < \infty, \tag{A.1.13}$$

$$\int_0^\infty \frac{u\sin ut\, J_1(bu)}{u^2 + K^2}\, du = K_1(Kb)\,\sinh Kt, \quad K > 0,\ 0 < t < b, \tag{A.1.14}$$

where $I_\nu(x)$ and $K_\nu(x)$ are the modified Bessel functions.

$$\int_0^\infty J_\nu(xt)\, J_\nu(yt)\, dt = \frac{2}{\pi(xy)^\nu}\int_0^{\min(x,y)} \frac{s^{2\nu}ds}{\sqrt{(x^2 - s^2)(y^2 - s^2)}}, \tag{A.1.15}$$

$$\int_0^\lambda t\, J_\nu(xt)\, J_\nu(yt)\, dt$$

$$= \frac{\lambda x\, J_{\nu+1}(\lambda x)\, J_\nu(\lambda y) - \lambda y\, J_{\nu+1}(\lambda y)\, J_\nu(\lambda x)}{x^2 - y^2} \tag{A.1.16}$$

$$\int_0^\infty t^{1-2n}\, \frac{J_\nu(at)\, J_\nu(bt)}{t^2 + c^2}\, dt$$

$$= \begin{cases} (-1)^n\, c^{-2n}\, I_\nu(bc)\, K_\nu(ac), & 0 < b < a,\ \mathrm{Re}\, c > 0,\ \mathrm{Re}\, \nu > n - 1, \\ (-1)^n\, c^{-2n}\, I_\nu(ac)\, K_\nu(bc), & 0 < a < c,\ \mathrm{Re}\, c > 0,\ \mathrm{Re}\, \nu > n - 1, \end{cases} \tag{A.1.17}$$

$$\int_0^\infty t^{\mu-\nu+1} J_\mu(bt) \, J_\nu(at) \, dt$$

$$= \begin{cases} \dfrac{(a^2-b^2)^{\nu-\mu-1} \, b^\mu}{2^{\nu-\mu-1} \, a^\nu \, \Gamma(\nu-\mu)}, & a > b, \\[3mm] 0, & a < b; \ Re \ \nu > Re \ \mu > -1. \end{cases} \tag{A.1.18}$$

The result (A.1.10) is due to Ursell (1947).

A.1.2 An integral involving Jacobi polynomials

We consider the integral

$$I = \int_0^1 x^{1+l} \, (1-x^2)^{-m} \, P_n^{l,-m} \, (1-2x^2) \, J_l(xy) \, dx, \tag{A.1.19}$$

where $P_n^{l,m}(x)$ is the Jacobi polynomial and $J_l(xy)$ is the Bessel function of the first kind of order l.

Using the well-known relation

$$P_n^{l,m}(x) = \frac{\Gamma(1+l+m)}{n! \, \Gamma(1+l)} \, F(-n, n+l+m+1; \, l+1; \frac{1-x}{2}), \tag{A.1.20}$$

where $F(a,b;c;z)$ is the hypergeometric function, the integral (A.1.19) becomes

$$I = \frac{\Gamma(1+l+m)}{n! \, \Gamma(1+l)} \int_0^1 x^{1+l}(1-x^2)^{-m} \, F(-n, n+l-m+1; \, l+1; x^2) \, J_l(xy) \, dx. \tag{A.1.21}$$

Utilizing the relation

$$F(a,b;c;z) = (1-z)^{c-b-a} \, F(c-a, c-b; c; z), \tag{A.1.22}$$

the representation (A.1.21) becomes

$$I = \frac{\Gamma(1+l+m)}{n! \, \Gamma(1+l)} \int_0^1 x^{1+l} F(m-n, 1+l+m; l+1; x^2) \, J_l(xy) \, dx. \tag{A.1.23}$$

Now, we have the relation (cf. Gradshteyn and Ryzhik 1992, p. 692)

$$\int_0^\infty J_\nu(at) \, J_\mu(bt) \, t^{-\lambda} \, dt = \frac{2^{-\lambda} \, b^\mu \, \Gamma\left(\frac{\nu+\mu-\lambda+1}{2}\right)}{a^{\mu-\lambda+1}\Gamma\left(\frac{\nu-\mu+\lambda+1}{2}\right) \Gamma(1+\mu)}$$

$$\times F\left(\frac{\nu+\mu-\lambda+1}{2}, \frac{-\nu+\mu-\lambda+1}{2}; 1+\mu; \frac{b^2}{a^2}\right), \tag{A.1.24}$$

where $Re(\nu + \mu - \lambda + 1) > 0$, $Re\ \lambda > -1$ and $0 < b < a$.

By the Hankel inversion theorem together with the relation

$$F(a, b; c; z) = F(b, a; c; z),$$

from (A.1.24), we get

$$\int_0^\infty b^{1+\mu} F\left(\frac{-\nu + \mu - \lambda + 1}{2}, \frac{\nu + \mu - \lambda + 2}{2}; 1 + \mu; \frac{b^2}{a^2}\right) J_\mu(bt)\ db$$

$$= \frac{2^\lambda \Gamma\left(\frac{\nu - \mu + \lambda + 1}{2}\right) \Gamma(1 + \mu)}{a^{\lambda - \mu - 1} \Gamma\left(\frac{\nu + \mu - \lambda + 1}{2}\right) t^{1+\lambda}}\ J_\nu(at). \tag{A.1.25}$$

From the relations (A.1.21) and (A.1.25), we deduce that

$$\int_0^1 x^{1+l}(1 - x^2)^{-m}\ P_n^{l,-m}(1 - 2x^2)\ J_l(xy)\ dx$$

$$= \frac{\Gamma(1 - l - m)}{2^m\ n!\ y^{1-m}}\ J_{1+l-m+2n}(y). \tag{A.1.26}$$

Using the Hankel operator defined by

$$S_{\beta,\alpha} f(x) = 2^\alpha x^{-\alpha} \int_0^\infty t^{1-\alpha} J_{2\beta+\alpha}(xt)\ f(t)\ dt,$$

the representation (A.1.26) can be expressed as

$$S_{l,-l}\left\{(1 - x^2)^{-m}\ P_n^{l,-m}(1 - 2x^2)\right\}$$

$$= \frac{\Gamma(1 - m + n)}{2^{l+m}\ n!\ x^{1-l+m}}\ J_{1+l-m+2n}(x). \tag{A.1.27}$$

From the relation (A.1.27), we find that

$$(1 - x^2)^{-m}\ P_n^{l,-m}(1 - 2x^2) = \frac{\Gamma(1 - m + n)}{2^{l+m}\ n!}\ S_{0,l}\left\{\frac{J_{1+l-m+2n}(2x)}{x^{1-l-m}}\right\}, \tag{A.1.28}$$

where use of the following identity is made

$$S_{\beta,\alpha}^{-1} = S_{\beta+\alpha,-\alpha} \tag{A.1.29}$$

for the inverse Hankel operator.

These results have been deduced by Rahman (1995).

A.1.3 Some discontinuous integrals involving Macdonald functions

Lebedev and Skal'skaya (1974) derived the following discontinuous integrals involving Macdonald functions.

Let us define

$$W^{\pm}(\lambda t, i\tau) = \frac{K_{\frac{1}{2}+i\tau}(\lambda t) \pm K_{\frac{1}{2}-i\tau}(\lambda t)}{2}.$$

Then

$$\left(\frac{2}{\pi}\right)^{\frac{3}{2}} \int_0^\infty \cosh\pi\tau\ W^+(\lambda t, i\tau)\ K_{i\tau}(\lambda r)\ d\tau$$

$$= \begin{cases} \dfrac{e^{-\lambda(t-r)}}{\sqrt{\lambda(t-r)}}, & r < t, \\ 0, & r > t. \end{cases} \tag{A.1.30}$$

$$\left(\frac{2}{\pi}\right)^{\frac{3}{2}} \int_0^\infty \tau\sinh\pi\tau\ \left\{\int_0^t W^+(\lambda s, i\tau)\ ds\right\}\ K_{i\tau}(\lambda r)\ d\tau$$

$$= \begin{cases} \dfrac{\sqrt{r}e^{-\lambda r}}{\sqrt{\lambda}} + \sqrt{\pi r}\ \Phi(\sqrt{\lambda r}), & r < t, \\[4mm] \dfrac{\sqrt{r}e^{-\lambda r}}{\sqrt{\lambda}} + \sqrt{\pi r}\ \Phi(\sqrt{\lambda r}) \\[2mm] \quad -\dfrac{r e^{-\lambda(r-t)}}{\sqrt{\lambda(r-t)}} - \sqrt{\pi r}\ \Phi\left(\sqrt{\lambda(r-t)}\right), & r > t, \end{cases} \tag{A.1.31}$$

where $\Phi(x) = \dfrac{2}{\sqrt{\pi}} \int_0^x e^{-t^2} dt$.

$$\left(\frac{2}{\pi}\right)^{\frac{3}{2}} \int_0^\infty \sinh\pi\tau\ W^-(\lambda t, i\tau)\ K_{i\tau}(\lambda r)\ d\tau$$

$$= \begin{cases} 0, & r < t, \\ \dfrac{e^{-\lambda(r-t)}}{\sqrt{\lambda(r-t)}}, & r > t. \end{cases} \tag{A.1.32}$$

203

$$\left(\frac{2}{\pi}\right)^{\frac{3}{2}} \int_0^\infty \tau \cosh \pi\tau \left\{\int_0^t W^-(\lambda s, i\tau) \, ds\right\} K_{i\tau}(\lambda r) \, d\tau$$

$$= \begin{cases} \dfrac{r \, e^{-\lambda(t-r)}}{\sqrt{\lambda(t-r)}} + \sqrt{\pi} r \, \Phi\left(\sqrt{\lambda(t-r)}\right), & r < t, \\ 0, & r > t. \end{cases} \qquad \text{(A.1.33)}$$

The relations (A.1.31)−(A.1.33) can be established by expanding their right sides into the Kontorovich−Lebedev integral expansion formula (cf. Lebedev 1965)

$$f(r) = \frac{2}{\pi^2} \int_0^\infty \tau \, \sinh \pi\tau \, K_{i\tau}(\lambda r)$$

$$\times \left\{\int_0^\infty \frac{f(t)}{t} K_{i\tau}(\lambda t) \, dt\right\} d\tau, \ 0 < r < \infty. \qquad \text{(A.1.34)}$$

The relation (A.1.30) can be established using the integral expansion

$$f(r) = \frac{2}{\pi^2} \int_0^\infty \tau \, \sinh \pi\tau \, K_{i\tau}(\lambda r)$$

$$\times \left\{\frac{\pi f(0)}{\tau \, \sinh \pi\tau} + \int_0^\infty \frac{[f(t) - f(0) \, e^{-\lambda t}]}{t} K_{i\tau}(\lambda t) \, dt\right\} d\tau, \ 0 < r < \infty, \qquad \text{(A.1.35)}$$

which generalizes the formula (A.1.34) when $f(r)$ tends to a non-zero limit as $r \to 0$.

A.1.4 Weber−Orr integral transforms

The Weber−Orr integral transform is suitable for use in a region outside a circle. This integral transform was first discovered formally by Weber (1873). Orr (1909) rediscovered this transform using a method based on the theory of complex variables. Nicholson (1921) first applied this transform to a heat conduction problem involving an infinite region with a cylindrical hole. Later, Titchmarsh (1923) gave a rigorous proof of the Weber−Orr integral expansion theorem in the following form.

If $x > a > 0$ and $\sqrt{t} f(t)$ is summable in the infinite interval (a, ∞) and $f(t)$ is of bounded variation in a neighbourhood of $t = x$, then at the point of continuity, for any real ν

$$f(x) = \int_0^\infty u \, \frac{R_\nu(xu, au)}{J_\nu^2(au) + Y_\nu^2(au)} \, F(u) \, du, \qquad \text{(A.1.36)}$$

where

$$F(u) = \int_a^\infty t \, R_\nu(tu, au) \, f(t) \, dt, \qquad (A.1.37)$$

with

$$R_\nu(x, y) \equiv J_\nu(x) \, Y_\nu(y) - Y_\nu(x) \, J_\nu(y).$$

Watson (1944) presented a rigorous proof of another form of Weber−Orr integral expansion formula given by

$$f(x) = \int_0^\infty \xi \, R_\nu(\xi x, ax) \, F(\xi) \, d\xi, \qquad (A.1.38)$$

where

$$F(\xi) = \int_0^\infty t \, \frac{R_\nu(\xi t, at)}{J_\nu^2(at) + Y_\nu^2(at)} \, f(t) \, dt. \qquad (A.1.39)$$

It may be noted here that if one defines the Weber−Orr transform in the following way

$$\overline{f}(\xi) = W_{\nu,\nu}\left[f(x); \xi\right] \equiv \int_a^\infty x \, R_{\nu,\nu}(\xi x, \xi a) \, f(x) \, dx, \qquad (A.1.40)$$

then its inverse transform takes the form

$$f(x) = W_{\nu,\nu}^{-1}\left[\overline{f}(\xi); x\right] \equiv \int_0^\infty \xi \, \frac{R_{\nu,\nu}(\xi x, \xi a)}{J_\nu^2(a\xi) + Y_\nu^2(a\xi)} \, \overline{f}(\xi) \, d\xi, \qquad (A.1.41)$$

where (ν, ν) is called the order of the operator $W_{\nu,\nu}[;]$ and

$$R_{\nu,\nu}(x, y) \equiv J_\nu(x) \, Y_\nu(y) - Y_\nu(x) \, J_\nu(y).$$

A.1.5 Associated Weber−Orr integral transforms

For applications in the theory of elasticity, thermoelasticity and other problems of mathematical physics, the aforesaid Weber−Orr integral transforms may not be sufficient to solve a problem having cylindrical symmetry. Depending on the boundary conditions at $r = a$, say, one has to employ some other kind of integral transforms which are closely related to Weber−Orr integral transforms and are called associated Weber−Orr integral transforms (cf. Krajewskii and Olesiak 1982).

205

To simplify the notation, let the operators $W_{\mu,\nu}[f(x);\xi]$ and $W_{\mu,\nu}^{-1}[F(\xi);x]$ (not necessarily the inverse of the operator $W_{\mu,\nu}$ for any μ) be defined by

$$W_{\mu,\nu}[f(x);\xi] \equiv \int_a^\infty x\, R_{\mu,\nu}(\xi x, \xi a)\, f(x)\, dx, \ a > 0, \qquad (A.1.42)$$

and

$$W_{\mu,\nu}^{-1}[F(\xi);x] \equiv \int_0^\infty \xi\, \frac{R_{\mu,\nu}(\xi x, \xi a)}{J_\nu^2(\xi a) + Y_\nu^2(\xi a)}\, F(\xi)\, d\xi, \qquad (A.1.43)$$

where

$$R_{\mu,\nu}(\xi x, \xi a) \equiv J_\mu(\xi x) Y_\nu(\xi a) - Y_\mu(\xi x) J_\nu(\xi a).$$

Krajewskii and Olesiak (1982) first introduced associated Weber$-$Orr integral transforms together with their inversions for $\mu = \nu - 1$ and $\mu = \nu - 2$ $(\nu \geq 1)$ in the notation of (A.1.42) and (A.1.43). Nasim (1989) generalized these by taking $\mu = \nu - \alpha$ or $\nu + \alpha$ where $\nu > -\frac{1}{2}$ and $0 < \alpha < \frac{\nu}{2} + \frac{3}{4}$ or $0 < \alpha < \nu + \frac{3}{2}$ for which the notation (A.1.43) no longer gives the inversion formula. However, as a special case, if $\alpha = k$ (a non negative integer) then he derived the associated Weber$-$Orr integral transform and the corresponding inversion as

$$F(\xi) = W_{\nu-k,\nu}[f(x);\xi], \qquad (A.1.44)$$

and

$$f(x) = W_{\nu-k,\nu}^{-1}[F(\xi);x], \qquad (A.1.45)$$

Also, Olesiak (1990) obtained the associated Weber$-$Orr transform together with the inversion formula for $\mu = \nu + 1$ as

$$\Psi(\xi) = W_{\nu+1,\nu}[\psi(x);\xi], \qquad (A.1.46)$$

and

$$\psi(x) = W_{\nu+1,\nu}^{-1}[\Psi(\xi);x]. \qquad (A.1.47)$$

He also applied this transform in some problems of thermoelasticity.

A-2 Legendre and associated Legendre functions

A.2.1 The Legendre function of the first kind $P_{-\frac{1}{2}+i\tau}(\cosh x)$ and of the second kind $Q_{-\frac{1}{2}+i\tau}(\cosh x)$ are two independent solutions of the ordinary differential equation

$$\frac{1}{\sinh x}\frac{d}{dx}\left(\sinh x\frac{du}{dx}\right)+\left(\frac{1}{4}+\tau^2\right)u=0, \tag{A.2.1}$$

The integral representation for $P_{-\frac{1}{2}+i\tau}(\cosh x)$ is (cf. Sneddon 1972, p. 381)

$$P_{-\frac{1}{2}+i\tau}(\cosh x)=\frac{\sqrt{2}}{\pi}\int_0^x\frac{\cos\tau t}{\sqrt{\cosh x-\cosh t}}\,dt. \tag{A.2.2}$$

By the use of the Fourier cosine inversion theorem, (A.2.2) gives

$$\int_0^\infty P_{-\frac{1}{2}+i\tau}(\cosh x)\cos\tau t\,d\tau=\frac{H(x-t)}{\sqrt{2(\cosh x-\cosh t)}}, \tag{A.2.3}$$

where $H(x)$ is the Heaviside unit function.

Using the representations (A.2.2) and (A.2.3), the following result can be established

$$\int_0^\infty P_{-\frac{1}{2}+i\tau}(\cosh x)P_{-\frac{1}{2}+i\tau}(\cosh y)\,d\tau$$
$$=\frac{1}{\pi}\int_0^{\min(x,y)}\frac{dt}{\sqrt{(\cosh x-\cosh t)(\cosh y-\cosh t)}}. \tag{A.2.4}$$

The following two discontinuous integrals involving Legendre functions of the first kind have been established by Lebedev and Skal'skaya (1969):

$$\cosh t\int_0^\infty\frac{\tau\,\tanh\pi\tau}{\omega_\mu(\tau)}F\left(\frac{1}{4}+\frac{\mu}{2}+\frac{i\tau}{2},\frac{1}{4}+\frac{\mu}{2}-\frac{i\tau}{2};\frac{1}{2};-\sinh^2 t\right)P_{-\frac{1}{2}+i\tau}(\cosh x)\,d\tau$$

$$=\begin{cases}\dfrac{\cosh^{1-\mu}t}{\sqrt{\cosh^2 x-\cosh^2 t}}F\left(-\dfrac{\mu}{2},\dfrac{\mu}{2};\dfrac{1}{2};\dfrac{\cosh^2 t-\cosh^2 x}{\cosh^2 t}\right), & t<x, \\[4mm] 0, & t>x,\end{cases} \tag{A.2.5}$$

for $0\le\mu\le\frac{1}{2}$, where

$$\omega_\mu(\tau)=4\pi^2\left\{\cosh\pi\tau\,\Gamma\left(\frac{1}{4}+\frac{\mu}{2}+\frac{i\tau}{2}\right)\Gamma\left(\frac{1}{4}+\frac{\mu}{2}-\frac{i\tau}{2}\right)\right.$$
$$\left.\times\Gamma\left(\frac{1}{4}-\frac{\mu}{2}+\frac{i\tau}{2}\right)\Gamma\left(\frac{1}{4}-\frac{\mu}{2}-\frac{i\tau}{2}\right)\right\}^{-1}$$

$$\sinh t \int_0^\infty \tau \tanh \pi\tau \; F\left(\frac{1}{4}+\frac{\mu}{2}+\frac{i\tau}{2}, \frac{1}{4}+\frac{\mu}{2}-\frac{i\tau}{2}; \frac{3}{2}; -\sinh^2 t\right) P_{-\frac{1}{2}+i\tau}(\cosh x) \; d\tau$$

$$= \begin{cases} 0, & t < x \\ \dfrac{\cosh^{1-\mu} t}{\sqrt{\cosh^2 t - \cosh^2 x}} F\left(\dfrac{1-\mu}{2}, \dfrac{\mu-1}{2}; \dfrac{1}{2}; \dfrac{\cosh^2 t - \cosh^2 x}{\cosh^2 t}\right), & t > x, \end{cases} \quad \text{(A.2.6)}$$

for $\mu \geq 0$

The integral representations (A.2.5) and (A.2.6) are new. The method of evaluation of the integrals on the left side of (A.2.5) and (A.2.6) are given in the paper of Lebedev and Skal'skaya (1969). The integral in (A.2.5) is also valid for $\mu > \frac{1}{2}$, but in this case the distribution of singular points in the complex ν-plane turns out to be somewhat different. This leads to the appearance of additional terms in the right side of (A.2.5). The number of these terms depends on the value of μ, which will be continuous functions of t in $(0, \infty)$.

A.2.2 The associated Legendre function of the first kind $P^\mu_{-\frac{1}{2}+i\tau}(\cosh x)$ and of the second kind $Q^\mu_{-\frac{1}{2}+i\tau}(\cosh x)$ satisfy the ordinary differential equation

$$\frac{1}{\sinh x} \frac{d}{dx}\left(\sinh x \frac{du}{dx}\right) + \left\{\left(\frac{1}{4}+\tau^2\right) - \frac{\mu^2}{\sinh^2 x}\right\} u = 0. \quad \text{(A.2.7)}$$

The following two integral representations (cf. Pathak 1978) for the associated Legendre function of the first kind $P^\mu_{-\frac{1}{2}+i\tau}(\cosh x)$ are useful to obtain some results involving $P^\mu_{-\frac{1}{2}+i\tau}(\cosh x)$:

$$P^\mu_{-\frac{1}{2}+i\tau}(\cosh x) = \sqrt{\frac{2}{\pi}} \frac{\sinh^\mu x}{\Gamma(\frac{1}{2}-\mu)} \int_0^x \frac{\cos \tau t}{(\cosh x - \cosh t)^{\frac{1}{2}+\mu}} \, dt, \quad \text{(A.2.8)}$$

where $Re\, \mu < \frac{1}{2}$, and

$$P^\mu_{-\frac{1}{2}+i\tau}(\cosh x) = \sqrt{2\pi} \, \frac{\{\Gamma(\frac{1}{2}+\mu)\}^{-1} \sinh^{-\mu} x \, \text{cosech } \pi\tau}{\Gamma(\frac{1}{2}-\mu+i\tau)\Gamma(\frac{1}{2}-\mu-i\tau)}$$

$$\times \int_x^\infty \frac{\sin \tau t}{(\cosh t - \cosh x)^{\frac{1}{2}-\mu}} \, dt, \quad \text{(A.2.9)}$$

where $|Re\, \mu| < \frac{1}{2}$.

By the Fourier cosine and sine inversion theorems, from (A.2.8) and (A.2.9) we get

$$\int_0^\infty P_{-\frac{1}{2}+i\tau}^\mu(\cosh x) \cos t\tau \, d\tau = \sqrt{\frac{\pi}{2}} \frac{\sinh^\mu x \, H(x-t)}{\Gamma(\frac{1}{2}-\mu)(\cosh x - \cosh t)^{\frac{1}{2}+\mu}}, \qquad (A.2.10)$$

and

$$\int_0^\infty \frac{1}{\pi} \Gamma\left(\frac{1}{2} - \mu + i\tau\right) \Gamma\left(\frac{1}{2} - \mu - i\tau\right) \sinh \pi\tau \, P_{-\frac{1}{2}+i\tau}^\mu(\cosh x) \sin t\tau \, d\tau$$

$$= \sqrt{\frac{\pi}{2}} \frac{\sinh^{-\mu} x \, H(t-x)}{\Gamma(\frac{1}{2}+\mu)(\cosh t - \cosh x)^{\frac{1}{2}-\mu}}. \qquad (A.2.11)$$

Using the above two integral representations (A.2.10) and (A.2.11), the following two results can easily be derived:

$$\int_0^\infty P_{-\frac{1}{2}+i\tau}^\mu(\cosh x) \, P_{-\frac{1}{2}+i\tau}^\mu(\cosh y) \, d\tau = \frac{(\sinh x \, \sinh y)^\mu}{\{\Gamma(\frac{1}{2}-\mu)\}^2}$$

$$\times \int_0^{\min(x,y)} \frac{dt}{\{(\cosh x - \cosh t)(\cosh y - \cosh t)\}^{\frac{1}{2}+\mu}}, \quad Re \, \mu < \frac{1}{2}, \qquad (A.2.12)$$

and

$$\int_0^\infty \left\{ \frac{\Gamma(\frac{1}{2} - \mu + i\tau)\Gamma(\frac{1}{2} - \mu - i\tau)\sinh \pi\tau}{\pi} \right\}^2 P_{-\frac{1}{2}+i\tau}^\mu(\cosh x) P_{-\frac{1}{2}+i\tau}^\mu(\cosh y) \, d\tau$$

$$= \frac{(\sinh x \, \sinh y)^{-\mu}}{\{\Gamma(\frac{1}{2}+\mu)\}^2} \int_{\max(x,y)}^\infty \frac{dt}{\{(\cosh t - \cosh x)(\cosh t - \cosh y)\}^{\frac{1}{2}-\mu}}, \quad |Re \, \mu| < \frac{1}{2}. \qquad (A.2.13)$$

Let $F_c(\tau)$ and $F_s(\tau)$ denote the Fourier cosine and sine transforms of $f(x)$. Then (A.2.8) can be written as

$$F_c\left[H(x-t)(\cosh x - \cosh t)^{-\frac{1}{2}-\mu}\right] = \Gamma\left(\frac{1}{2} - \mu\right) \sinh^{-\mu} x \, P_{-\frac{1}{2}+i\tau}^\mu(\cosh x). \quad (A.2.14)$$

By the Fourier cosine inversion theorem, (A.2.14) produces

$$\frac{H(x-t)}{(\cosh x - \cosh t)^{\frac{1}{2}+\mu}} = \sqrt{\frac{2}{\pi}} \, \Gamma\left(\frac{1}{2} - \mu\right) \sinh^{-\mu} x \int_0^\infty P_{-\frac{1}{2}+i\tau}^\mu(\cosh x) \cos \tau t \, d\tau, \qquad (A.2.15)$$

which is the relation (A.2.10).

Also, the representation (A.2.9) can be expressed as

$$F_s \left[H(t - x)(\cosh t - \cosh x)^{-\frac{1}{2}+\mu} \right] = \frac{\Gamma(\frac{1}{2} + \mu)\Gamma(\frac{1}{2} - \mu + i\tau)\Gamma(\frac{1}{2} - \mu - i\tau)}{\pi}$$

$$\times \sinh \pi\tau \, \sinh^\mu x \, P^\mu_{-\frac{1}{2}+i\tau}(\cosh x), \qquad (A.2.16)$$

which on inversion produces

$$\frac{H(t - x)}{(\cosh t - \cosh x)^{-\frac{1}{2}+\mu}} = \frac{\sqrt{2}}{\pi\sqrt{\pi}} \, \Gamma\left(\frac{1}{2} + \mu\right) \sinh^\mu x$$

$$\times \int_0^\infty \Gamma\left(\frac{1}{2} - \mu + i\tau\right) \Gamma\left(\frac{1}{2} - \mu - i\tau\right) \sinh \pi\tau \, P^\mu_{-\frac{1}{2}+i\tau}(\cosh x) \, \sin t\tau \, d\tau; \quad (A.2.17)$$

this gives the relation (A.2.11).

Equation (A.2.8) is an Abel type integral equation of the first kind and hence by Abel's inversion theorem, we get

$$\cos t\tau = \frac{\Gamma(\frac{1}{2} - \mu) \cos \pi\mu}{\sqrt{2\pi}} \frac{d}{dt} \int_0^t \frac{\sinh^{1-\mu} x \, P^\mu_{-\frac{1}{2}+i\tau}(\cosh x)}{(\cosh t - \cosh x)^{\frac{1}{2}-\mu}} \, dx. \qquad (A.2.18)$$

Integrating the above relation with respect to t from 0 to t, we obtain

$$\sin t\tau = -\frac{\Gamma(\frac{1}{2} - \mu) \, \tau \, \cos \pi\mu}{\sqrt{2\pi}} \int_0^t \frac{\sinh^{1-\mu} x \, P^\mu_{-\frac{1}{2}+i\tau}(\cosh x)}{(\cosh t - \cosh x)^{\frac{1}{2}-\mu}} \, dx. \qquad (A.2.19)$$

In a similar way, the integral representation (A.2.9) gives

$$\sin t\tau = -\frac{1}{\pi\sqrt{2\pi}} \, \Gamma\left(\frac{1}{2} + \mu\right) \cos \pi\mu \, \Gamma\left(\frac{1}{2} - \mu + i\tau\right) \Gamma\left(\frac{1}{2} - \mu - i\tau\right) \sinh \pi\tau$$

$$\times \frac{d}{dt} \int_t^\infty \frac{\sinh^{1+\mu} x \, P^\mu_{-\frac{1}{2}+i\tau}(\cosh x)}{(\cosh x - \cosh t)^{\frac{1}{2}+\mu}} dx. \qquad (A.2.20)$$

From the relation (A.2.15), by the generalized Mehler–Fock inversion theorem (cf. Mandal and Mandal 1997, p. 47), we find that

$$\cos t\tau = 2^{-\frac{1}{2}}\pi^{-\frac{3}{2}} \, \Gamma\left(\frac{1}{2} + \mu\right) \cos \pi\mu \, \Gamma\left(\frac{1}{2} - \mu + i\tau\right) \Gamma\left(\frac{1}{2} - \mu - i\tau\right) \tau \, \sinh \pi\tau$$

$$\times \int_t^\infty \frac{\sinh^{1+\mu} x \, P^\mu_{-\frac{1}{2}+i\tau}(\cosh x)}{(\cosh x - \cosh t)^{\frac{1}{2}+\mu}} \, dx. \qquad (A.2.21)$$

A-3 Generalized associated Legendre functions

One of the two linearly independent solutions of the generalized associated Legendre equation (cf. Mandal and Mandal 1997, p. 16)

$$(1 - z^2)\frac{d^2u}{dz^2} - 2z\,\frac{du}{dz} + \left\{ k(k+1) - \frac{\mu^2}{2(1-z)} - \frac{\nu^2}{2(1+z)} \right\} u = 0, \qquad (A.3.1)$$

is $P_k^{\mu,\nu}(z)$. This is called the generalized associated Legendre function of the first kind.

Using the relation (cf. Erdélyi et al. 1953)

$$F(a,b;c;z) = \frac{\Gamma(c)}{\Gamma(\lambda)\Gamma(c-\lambda)} \int_0^1 t^{\lambda-1}(1-t)^{c-\lambda-1}(1+tz)^{-d}$$

$$\times F(a-d,b;\lambda;tz)\, F\left(d, b-\lambda; c-\lambda; \frac{z(1-t)}{1-tz}\right) dt,$$

for $Re\ c > Re\lambda > 0$, $|arg(1-z)| < \pi$, the integral representation for $P_{-\frac{1}{2}+i\tau}^{\mu,\nu}(\cosh x)$ can be established as (cf. Virchenko 1984)

$$P_{-\frac{1}{2}+i\tau}^{\mu,\nu}(\cosh x) = \frac{2^{\frac{\nu-\mu+1}{2}}\sinh^\mu x}{\sqrt{\pi}\,\Gamma(\frac{1}{2}-\mu)} \int_0^x (\cosh x - \cosh t)^{-\mu-\frac{1}{2}}$$

$$\times F\left(\frac{\nu-\mu}{2}, -\frac{\nu+\mu}{2}; \frac{1}{2}-\mu; \frac{\cosh x - \cosh t}{1+\cosh x}\right\} \cos t\tau\ dt, \qquad (A.3.2)$$

where $Re\ \mu < \frac{1}{2}, |Re\ \nu| < 1 - Re\ \mu$.

Applying the inversion of an Abel type integral equation of the first kind

$$\int_a^x g(t)\,[\phi(x) - \phi(t)]^{-m}\, F\left(a,b;c; \frac{\phi(x) - \psi(t)}{1+\phi(x)}\right) dt = h(x), \qquad (A.3.3)$$

where $m = \frac{1}{2} + \mu$, $a = \frac{\nu-\mu}{2}$, $b = -\frac{\nu+\mu}{2}$, $C = \frac{1}{2} - \mu$, $\phi(x) = \cosh x$ and $\psi(t) = \cosh t$, as

$$g(x) = \frac{1}{\Gamma(\frac{1}{2}-\mu)\Gamma(\frac{1}{2}+\mu)} \frac{d}{dx}\left\{ (\phi(x)+1)^{\frac{\nu-\mu}{2}} \int_a^x [\phi(x) - \psi(t)]^{\mu-\frac{1}{2}}\,(1+\phi(t))^{\frac{\mu-\nu}{2}} \right.$$

$$\left. \times F\left(\frac{\mu-\nu}{2}, \frac{1+\mu-\nu}{2}; \frac{1}{2}+\mu; \frac{\phi(x)-\psi(t)}{1+\phi(x)}\right) h(t)\,\phi'(t)\,dt \right\}, \qquad (A.3.4)$$

from (A.3.2), we get

$$\cos t\tau = \frac{\Gamma(\frac{1}{2} - \mu)\cos \pi\mu}{\sqrt{\pi}2^{(\nu-\mu+1)/2}} \frac{d}{dt}\left[(\cosh t + 1)^{\frac{\nu-\mu}{2}} \int_0^t (\cosh t - \cosh x)^{\mu-\frac{1}{2}}(\cosh x + 1)^{\frac{\mu-\nu}{2}}\right.$$

$$\left.\times F\left(\frac{\mu-\nu}{2}, \frac{1+\mu-\nu}{2}; \frac{1}{2} + \mu; \frac{\cosh t - \cosh x}{1 + \cosh t}\right) P_{-\frac{1}{2}+i\tau}^{\mu,\nu}(\cosh x)\ \sinh^{1-\mu} x\ dx\right].$$

$$(A.3.5)$$

Let

$$f^*(\tau) = \int_0^\infty f(x)\ P_{-\frac{1}{2}+i\tau}^{\mu,\nu}(\cosh x)\ \sinh x\ dx, \qquad (A.3.6)$$

be the generalized Mehler−Fock integral transform of the function $f(x)$. Then under certain conditions (cf. Mandal and Mandal 1997, p. 57, Theorem 2.4.5)

$$f(x) = \frac{2^{\mu-\nu-1}}{\pi^2} \int_0^\infty \tau\ \sinh 2\pi\tau\ \Gamma\left(\frac{1-\mu+\nu}{2} + i\tau\right)\Gamma\left(\frac{1-\mu+\nu}{2} - i\tau\right)$$

$$\times \Gamma\left(\frac{1-\mu-\nu}{2} + i\tau\right)\Gamma\left(\frac{1-\mu-\nu}{2} - i\tau\right) P_{-\frac{1}{2}+i\tau}^{\mu,\nu}(\cosh x)\ f^*(\tau)\ d\tau. \qquad (A.3.7)$$

Let $F_c(\tau)$ denote the Fourier cosine transform of $f(x)$. Then (A.3.2) can be written as

$$F_c\left[\frac{H(x-t)}{(\cosh x - \cosh t)^{\frac{1}{2}+\mu}}F\left(\frac{\nu-\mu}{2}, -\frac{\nu+\mu}{2}; \frac{1}{2} - \mu; \frac{\cosh x - \cosh t}{1 + \cosh x}\right)\right]$$

$$= 2^{\frac{\mu-\nu}{2}}\Gamma\left(\frac{1}{2} - \mu\right)\sinh^{1-\mu} x\ P_{-\frac{1}{2}+i\tau}^{\mu,\nu}(\cosh x), \qquad (A.3.8)$$

where $H(x)$ is the Heaviside unit function, so that by the Fourier cosine inversion theorem, we obtain

$$\frac{H(x-t)}{(\cosh x - \cosh t)^{\frac{1}{2}+\mu}}F\left(\frac{\nu-\mu}{2}, -\frac{\nu+\mu}{2}; \frac{1}{2} - \mu; \frac{\cosh x - \cosh t}{1 + \cosh x}\right)$$

$$= \pi^{-\frac{1}{2}}2^{\frac{\mu-\nu+1}{2}}\Gamma\left(\frac{1}{2} - \mu\right)\sinh^{-\mu} t \int_0^\infty P_{-\frac{1}{2}+i\tau}^{\mu,\nu}(\cosh x)\cos t\tau\ d\tau. \qquad (A.3.9)$$

The above equation can be written as

$$\int_0^\infty \frac{2^{\mu-\nu-1} \tau \, \sinh 2\pi\tau}{\pi^2} \Gamma\left(\frac{1-\mu+\nu}{2}+i\tau\right) \Gamma\left(\frac{1-\mu+\nu}{2}-i\tau\right)$$

$$\times \Gamma\left(\frac{1-\mu-\nu}{2}+i\tau\right) \Gamma\left(\frac{1-\mu-\nu}{2}-i\tau\right) P^{\mu,\nu}_{-\frac{1}{2}+i\tau}(\cosh x) \, g^*(\tau) \, d\tau$$

$$= \pi^{-\frac{1}{2}} 2^{\frac{\nu-\mu-1}{2}} \left\{\Gamma\left(\frac{1}{2}-\mu\right)\right\}^{-1} \sinh^\mu t \, (\cosh x - \cosh t)^{-\frac{1}{2}-\mu}$$

$$\times F\left(\frac{\nu-\mu}{2}; -\frac{\nu+\mu}{2}; \frac{1}{2}-\mu; \frac{\cosh x - \cosh t}{1+\cosh x}\right) H(x-t),$$

where

$$g^*(\tau) = \pi^2 \, 2^{\nu-\mu+1} \tau^{-1} \, \sinh^{-1} 2\pi\tau \, \cos t\tau \, \left\{\Gamma\left(\frac{1-\mu+\nu}{2}+i\tau\right)\right.$$

$$\times \left. \Gamma\left(\frac{1-\mu+\nu}{2}-i\tau\right) \Gamma\left(\frac{1-\mu-\nu}{2}+i\tau\right) \Gamma\left(\frac{1-\mu-\nu}{2}-i\tau\right)\right\}^{-1}.$$

Now comparing the above relation with (A.3.7) and then by the generalized Mehler −Fock transform formula (A.3.6), we get

$$\cos t\tau = \pi^{-3/2} 2^{\frac{\mu-\nu-3}{2}} \left\{\Gamma(\frac{1}{2}-\mu)\right\}^{-1} \tau \, \sinh 2\pi\tau \, \Gamma\left(\frac{1-\mu+\nu}{2}+i\tau\right)$$

$$\times \Gamma\left(\frac{1-\mu+\nu}{2}-i\tau\right) \Gamma\left(\frac{1-\mu-\nu}{2}+i\tau\right) \Gamma\left(\frac{1-\mu-\nu}{2}-i\tau\right)$$

$$\times \int_t^\infty (\cosh x - \cosh t)^{-\frac{1}{2}-\mu} F\left(\frac{\nu-\mu}{2}, -\frac{\nu+\mu}{2}; \frac{1}{2}-\mu; \frac{\cosh x - \cosh t}{1+\cosh x}\right)$$

$$\times P^{\mu,\nu}_{-\frac{1}{2}+i\tau}(\cosh x) \sinh^{1+\mu} x \, dx. \qquad \text{(A.3.10)}$$

Differentiating both sides of (A.3.10) with respect to t, we obtain

$$\sin t\tau = \pi^{-3/2} 2^{\frac{\mu-\nu-3}{2}} \cos \pi\mu \, \sinh 2\pi\tau \, \Gamma\left(\frac{1}{2}+\mu\right) \Gamma\left(\frac{1-\mu+\nu}{2}+i\tau\right)$$

$$\times \Gamma\left(\frac{1-\mu+\nu}{2}-i\tau\right) \Gamma\left(\frac{1-\mu-\nu}{2}+i\tau\right) \Gamma\left(\frac{1-\mu-\nu}{2}-i\tau\right)$$

$$\times \frac{d}{dt} \int_t^\infty \frac{F\left(\frac{\nu-\mu}{2}, -\frac{\nu+\mu}{2}; \frac{1}{2}-\mu; \frac{\cosh x - \cosh t}{1+\cosh x}\right)}{(\cosh x - \cosh t)^{\mu+\frac{1}{2}}}$$

$$\times P^{\mu,\nu}_{-\frac{1}{2}+i\tau}(\cosh x) \sinh^{1-\mu} x \, dx. \qquad \text{(A.3.11)}$$

Integrating both sides of (A.3.5) with respect to t from 0 to t, we get

$$\sin t\tau = \pi^{-\frac{1}{2}}2^{(\mu-\nu-1)/2}\ \Gamma\left(\frac{1}{2}-\mu\right)\ \tau\ \cos\pi\mu\ (\cosh t+1)^{\frac{\nu-\mu}{2}}$$

$$\times\int_0^t(\cosh t-\cosh x)^{\mu-\frac{1}{2}}(\cosh x+1)^{\frac{\mu-\nu}{2}}$$

$$\times F\left(\frac{\mu-\nu}{2},\frac{1+\mu-\nu}{2};\frac{1}{2}+\mu;\frac{\cosh t-\cosh x}{1+\cosh t}\right)$$

$$\times P^{\mu,\nu}_{-\frac{1}{2}+i\tau}(\cosh x)\sinh^{1-\mu}x\ dx. \tag{A.3.12}$$

The above equation can be written as

$$\pi^{\frac{1}{2}}2^{(\nu-\mu+1)/2}\left\{\Gamma\left(\frac{1}{2}-\mu\right)\right\}^{-1}\tau^{-1}\sec\pi\mu\ (\cosh t+1)^{(\mu-\nu)/2}\sin t\tau$$

$$=\int_0^\infty(\cosh t-\cosh x)^{\mu-\frac{1}{2}}(\cosh t+1)^{(\mu-\nu)/2}\ \sinh^{1-\mu}\ x$$

$$\times F\left(\frac{\mu-\nu}{2},\frac{1+\mu-\nu}{2};\frac{1}{2}+\mu;\frac{\cosh t-\cosh x}{1+\cosh t}\right)\ P^{\mu,\nu}_{-\frac{1}{2}+i\tau}(\cosh x)\ H(t-x)\ dx.$$

Now comparing the above equation with (A.3.6) and then by utilizing the inverse generalized associated Mehler−Fock transform formula (A.3.7), we have

$$(\cosh t-\cosh x)^{\mu-\frac{1}{2}}(\cosh x+1)^{\frac{\mu-\nu}{2}}\ F\left(\frac{\mu-\nu}{2},\frac{1+\mu-\nu}{2};\frac{1}{2}+\mu;\frac{\cosh t-\cosh x}{1+\cosh t}\right)$$

$$\times\sinh^{1-\mu}x\ H(t-x)=\pi^{-\frac{3}{2}}\ 2^{(\mu-\nu-1)/2}\left\{\Gamma\left(\frac{1}{2}-\mu\right)\right\}^{-1}\ \sec\pi\mu\ (\cosh t+1)^{\frac{\mu-\nu}{2}}$$

$$\times\int_0^\infty\sinh 2\pi\tau\ \Gamma\left(\frac{1-\mu+\nu}{2}+i\tau\right)\Gamma\left(\frac{1-\mu+\nu}{2}-i\tau\right)\Gamma\left(\frac{1-\mu-\nu}{2}+i\tau\right)$$

$$\times\Gamma\left(\frac{1-\mu-\nu}{2}-i\tau\right)P^{\mu,\nu}_{-\frac{1}{2}+i\tau}(\cosh x)\ \sin t\tau\ d\tau. \tag{A.3.13}$$

By Fourier sine inversion, the above equation becomes

$$P^{\mu,\nu}_{-\frac{1}{2}+i\tau}(\cosh x) = (2\pi)^{3/2}2^{\frac{\nu-\mu}{2}}\left\{\Gamma\left(\frac{1}{2}+\mu\right)\right\}^{-1} \text{cosech } 2\pi\tau \; \sinh^{-\mu} x$$

$$\times\left\{\Gamma\left(\frac{1-\mu+\nu}{2}+i\tau\right)\Gamma\left(\frac{1-\mu+\nu}{2}-i\tau\right)\right.$$

$$\times\Gamma\left(\frac{1-\mu-\nu}{2}+i\tau\right)\Gamma\left(\frac{1-\mu-\nu}{2}-i\tau\right)\right\}^{-1}$$

$$\times\int_x^\infty \frac{(\cosh t + 1)^{(\nu-\mu)/2}(\cosh t - \cosh x)^{-\frac{1}{2}+\mu}}{(\cosh t + 1)^{(\nu-\mu)/2}}$$

$$\times F\left(\frac{\mu-\nu}{2},\frac{1+\mu-\nu}{2};\frac{1}{2}+\mu;\frac{\cosh t - \cosh x}{1+\cosh t}\right)\sin t\tau \; dt.$$

Using the relation

$$F(a,b;c;z) = (1-z)^{-a}\; F\left(a,b;c;\frac{z}{z-1}\right),$$

the above equation gives the following integral representation for $P^{\mu,\nu}_{-\frac{1}{2}+i\tau}(\cosh x)$:

$$P^{\mu,\nu}_{-\frac{1}{2}+i\tau}(\cosh x) = (2\pi)^{3/2}2^{\frac{\nu-\mu}{2}}\left\{\Gamma\left(\frac{1}{2}+\mu\right)\right\}^{-1} \text{cosech } 2\pi\tau \; \sinh^{-\mu} x$$

$$\times\left\{\Gamma\left(\frac{1-\mu+\nu}{2}+i\tau\right)\Gamma\left(\frac{1-\mu+\nu}{2}-i\tau\right)\Gamma\left(\frac{1-\mu-\nu}{2}+i\tau\right)\right.$$

$$\times\Gamma\left(\frac{1-\mu-\nu}{2}-i\tau\right)\right\}^{-1}\int_x^\infty (\cosh t - \cosh x)^{-\frac{1}{2}+\mu}$$

$$\times F\left(\frac{\mu-\nu}{2},\frac{\mu+\nu}{2};\frac{1}{2}+\mu;\frac{\cosh x - \cosh t}{1+\cosh x}\right)\sin t\tau \; dt, \tag{A.3.14}$$

where $|Re\;\mu| < \frac{1}{2}$, $|Re\;\nu| < 1 - Re\;\mu$.

For $\mu = \nu$, all the above results reduce to the results given in A-2.

Bibliography

Aggarwala, B D and Nasim, C (1996), Steady-state temperature in a quarter plane. *Internat. J. Math. & Math. Sci.* **19**, pp. 371−380.

Aleksandrov, V M (1975), On a method of reducing dual integral equations and dual series equations to infinite algebraic systems. *PMM* **39**, pp. 303−311 (Engl. Transl.).

Aleksandrov, V M and Chebakov, M I (1973), On a method of solving dual integral equations. *PMM* **37**, pp. 1031−1041 (Engl. Transl.).

Anderssen, R S, De Hoog, F R and Rose, L R F (1982), Explicit solutions for a class of dual integral equations. *Proc. Roy. Soc. Edin.* **A91**, pp. 277−285.

Babloian, A A (1964), Solutions of certain dual integral equations. *PMM* **28**, pp. 1227−1236 (Engl. Transl.).

Banerjea, Sudeshna and Mandal, B N (1998), Scattering of water waves by a submerged thin vertical wall with a gapp. *J. Austral. Math. Soc. Ser. B.* **39**, pp. 318−331.

Busbridge, I W (1938), Dual integral equations. *Proc. Lond. Math. Soc.* **44**, pp. 115−129.

Chakrabarti, A (1989), On some dual integral equations involving Bessel functions of order one. *Indian J. Pure Appl. Math.* **20**, pp. 483−492.

Chakrabarti, A, Banerjea, S, Mandal, B N and Sahoo, T (1997), A unified approach to problems of scattering of surface water waves by vertical barriers. *J. Austral. Math. Soc. Ser. B.* **39**, pp. 93−103.

Chakrabarti, A, Mandal, B N, Banerjea, S and Sahoo, T (1995), Solution of a class of mixed boundary value problems for Laplace's equation arising in water wave scattering. *J. Indian Inst. Sci.* **75**, pp. 577–585.

Chakrabarti, A and Mandal, Nanigopal (1998), Solutions of some dual integral equations. *Z. angew Math. Mech.* **78**, pp. 141–144.

Cooke, J C (1970), The solution of some integral equations and their connection with dual integral equations and series. *Glasgow Math. J.* **11**, pp. 9–20.

Davis, A M J (1991), A translating disc in a Sampson flow; pressure driven flow through concentric holes in parallel walls. *Quart. Mech. Appl. Math.* **44**, pp. 471–486.

Dhaliwal, R S and Singh, B M (1987), Dual integral equations involving Legendre function of complex index. *J. Math. Phys. Sci.* **21**, pp. 307–313.

Erdélyi, A (1968), Some dual integral equations. *SIAM J. Appl. Math.* **16**, pp. 1338–1340.

Erdélyi, A, Magnus, W, Oberhittinger, F and Tricomi, F G (1953), *Higher Transcendental Functions,* Vol. 1. McGraw-Hill, New York.

Erdélyi, A, Magnus, W, Oberhittinger, F and Tricomi, F G (1954a), *Tables of Integral Transforms,* Vol. 1. McGraw-Hill, New York.

Erdélyi, A, Magnus, W, Oberhittinger, F and Tricomi, F G (1954b), *Tables of Integral Transforms,* Vol. 2. McGraw-Hill, New York.

Eswaran, K (1990), On the solutions of a class of dual integral equations occurring in diffraction problems. *Proc. Roy. Soc. Lond.* **A429**, pp. 399–427.

Gladwell, G M L and Low, R D (1970), On an initial value Reissner– Sagoci problem. *Int. J. Engng. Sci.* **8**, pp. 447–456.

217

Goodrich, F C (1969), The theory of absolute surface shear viscosity I. *Proc. Roy. Soc. Lond.* **A310**, pp. 359−372.

Gradshteyn, I S and Ryzhik, I M (1992), *Tables of Integrals, Series and Products.* Academic Press, New York.

Hobson, E W (1931), *Theory of Spherical and Ellipsoidal Harmonics*, Cambridge University Press, London.

Krajewskii, J and Olesiak, Z S (1982), Associated Weber−Orr integral transforms of $W_{\nu-1,\nu}[;]$ and $W_{\nu-2,\nu}[;]$ types. *Bull. de l'Acad. Polon. des Sci.* **30**, pp. 31−37.

Lebedev, N N (1965), *Special functions and their applications*, Prentice-Hall, New Jersey.

Lebedev, N N and Skal'skaya, I P (1969), The solution of one class of dual integral equations connected with the Mehler−Fock transform in the theory of elasticity and mathematical physics. *PMM* **33**, pp. 1061−1068 (Engl. Transl.).

Lebedev, N N and Skal'skaya, I P (1974), Dual integral equations related to Kontorovich−Lebedev transform. *PMM* **38**, pp. 1038−1040 (Engl. Transl.).

Love, E R (1967), Some integral equations involving Hypergeometric functions. *Proc. Edin. Math. Soc. Ser.* 2 **15**, pp. 169−198.

Mandal, B N (1988), A note on Bessel function dual integral equation with weight function. *Internat. J. Math. & Math. Sci.* **11**, pp. 543−550.

Mandal, B N (1992), A note on dual integral equations involving associated Legendre function. *Internat. J. Math. & Math. Sci.* **15**, pp. 601−604.

Mandal, B N, Banerjea, S and Kanoria, M (1997), The rolling ship problem −revisited. *Math. Comput. Modelling.* **25**, pp. 11−18.

Mandal, B N and Mandal, Nanigopal (1997), *Integral expansions related to Mehler−Fock type transforms, Pitman Research Notes in Mathematics Series*, Vol. 367. Addison Wesley Longman, UK.

Mandal, Nanigopal (1995), On a class of dual integral equations involving generalized associated Legendre functions. *Indian J. Pure Appl. Math.* **26**, pp. 1191−1204.

Mandal, Nanigopal and Mandal, B N (1993), On some dual integral equations involving Legendre and associated Legendre functions. *J. Indian Inst. Sci.* **73**, pp. 565−578.

Mandal, Nanigopal and Mandal, B N (1996), A note on dual integral equations involving inverse associated Weber−Orr transforms. *Internat. J. Math. & Math. Sci.* **19**, pp. 161−170.

Muskhelishvili, N I (1963), *Singular Integral Equations.* Noordhoff, Groningen.

Nasim, C (1986), On dual integral equations with Hankel kernel and an arbitrary weight function. *Internat. J. Math. & Math. Sci.* **9**, pp. 293−300.

Nasim, C (1989), Associated Weber integral transforms of arbitrary orders. *Indian J. Pure Appl. Math.* **20**, pp. 1126−1138.

Nasim, C (1991), Dual integral equations involving Weber−Orr transforms. *Internat. J. Math. & Math. Sci.* **14**, pp. 163−176.

Nasim, C and Aggarwala, B D (1984), On some dual integral equations. *Indian J. Pure Appl. Math.* **15**, pp. 323−340.

Nicholson, J W (1921), A problem in the theory of heat conduction. *Proc. Roy. Soc. Lond.* Series C. **A98**, p. 226.

Noble, B (1958), Certain dual integral equations. *J. Math. and Phys.* **37**, pp. 128−136.

Noble, B (1963), The solution of Bessel function dual integral equations by a multiplying factor method. *Proc. Camb. Phil. Soc.* **59**, pp. 351−362.

Olesiak, Z S (1990), Applications of Weber−Orr integral transforms in problems of thermoelasticity. *Elasticity, Mathematical Methods and Applications* (ed. by Eason, G. and Ogden, R. W.), I. N. Sneddon 70th Birthday Volume. Ellis Horwood, New York.

Orr, W M (1909), Extensions of Fourier's and the Bessel Fourier theorems. *Proc. Roy. Irish Acad.* **27**, pp. 205−248.

Pathak, R S (1978), On a class of dual integral equations. *Proc. Kon. Ned. Ak. V. Wet.* Amsterdam. **81**, pp. 491−501.

Protsenko, V S and Solov'ev, A I (1982), Certain hybrid integral transforms and their applications in the theory of elasticity of non-homogeneous media. *Prikl. Mekh.* **18**, pp. 62−67 (in Russian).

Rahman, Mujibur (1995), A note on the polynomial solution of a class of dual integral equations arising in mixed boundary value problems of elasticity. *Z. angew Math. Phys.* **46**, pp. 107−121.

Rose, L R F and De Hoog, F R (1983), Exact solutions of certain dual integral equations and their asymptotic properties. *Quart. J. Mech. Appl. Math.* **36**, pp. 419−436.

Rukhovets, A N and Ufliand, Ia S (1966), On a class of dual integral equations and their applications to the theory of elasticity. *PMM* **30**, pp. 334−341 (English Transl.).

Shail, R (1970), The impulsive Reissner−Sagoci problem. *J. Math. Mech.* **19**, pp. 709−716.

Shail, R (1978), The torque on a rotating disc in the surface of a liquid with an adsorbed film. *J. Engng. Math.* **12**, pp. 59−76.

Shail, R (1979), The slow rotation of an axisymmetric solid submerged in a fluid with a surfactant surface layer I − The rotating disc in a semi-infinite fluid. *Int. J. Multiphase Flow.* **5**, pp. 169−183.

Sneddon, I N (1960), The elementary solution of dual integral equations. *Proc. Glasgow Math. Assoc.* **4**, p. 108.

Sneddon, I N (1966), *Mixed Boundary Value Problems in Potential Theory.* North Holland Publ., Amsterdam.

Sneddon, I N (1972), *The Use of Integral Transforms.* McGraw-Hill, New York.

Sneddon, I N and Lowengrub, M (1969), *Crack Problems in the Mathematical Theory of Elasticity.* John Wiley & Sons, New York.

Srivastava, N (1990), Some theorems concerning dual and triple integral equations. *J. Maulana Azad College Tech.* **23**, pp. 77−82.

Srivastava, N (1991), On dual integral equations associated with Mehler−Fock transform. *J. Indian Acad. Math.* **13**, pp. 115−122.

Srivastava, N and Srivastava, K N (1991), On some dual integral equations. *Z. angew Math. Mech.* **71**, pp. 126−129.

Srivastav, R P (1964a), An axisymmetric mixed boundary value problem for a half-space with a cylindrical cavity. *J. Math. Mech.* **13**, pp. 385−393.

Srivastav, R P (1964b), A pair of dual integral equations involving Bessel functions of the first and second kind. *Proc. Edin. Math. Soc.* **14**, pp. 149−158.

Srivastav, R P (1965), Certain two-dimensional mixed boundary value problems for wedge-shaped regions and dual integral euqations. *Proc. Edin. Math. Soc.* **14**, pp. 321−332.

Srivastav, R P and Narain, P (1965), Certain two-dimensional problems of stress distribution in wedge-shaped elastic solids under discontinuous load. *Proc. Camb. Phil. Soc.* **61**, pp. 945−954.

Srivastav, R P and Parihar, K S (1968), Dual and triple integral equations involving inverse Mellin transforms. *SIAM J. Appl. Math.* **16**, pp. 126−133.

Titchmarsh, E C (1923), Weber's integral theorem, *Proc. Lond. Math. Soc.* **22**, pp. 15−28.

Titchmarsh, E C (1937), *Theory of Fourier Integrals*. Oxford University Press, London.

Ursell, F (1947), The effect of a fixed vertical barrier on surface waves in deep water. *Proc. Camb. Phil. Soc.* **43**, pp. 374−382.

Virchenko, N A (1984), Certain hybrid dual integral equations. *Ukrainian Mathematical Journal.* **36**, pp. 125−128.

Virchenko, N A (1989), *Dual (Triple) Integral Equations* (in Russian). Kiev University, Kiev.

Watson, G N (1944), *A Treatise on the Theory of Bessel Functions*. Cambridge University Press, Cambridge.

Weber, H (1873), Uber eine Darstelling Willkürlicher Functionen durch Besselśhe Functionen. *Math. Annal.* **6**, pp. 146−161.

Zlatina, I N (1972), Application of dual integral equations to the problem of torsion of an elastic space, weakened by a conical crack of finite dimensions. *PMM* **36**, pp. 1062−1068 (Engl. Transl.).

Index